Delmar's Test Preparation Series

Automobile Test

Automotive Engine Performance (Test A8)

2nd Edition

DELMAR
THOMSON LEARNING ™

Australia Canada Mexico Singapore Spain United Kingdom United States

DELMAR
THOMSON LEARNING™

Delmar's Test Preparation Series
Automobile Test

Automotive Engine Performance (Test A8)
2nd Edition

Delmar Staff:

Business Unit Director:
Alar Elken
Executive Editor:
Sandy Clark
Acquisitions Editor:
Jack Erjavec
Team Assistant:
Bryan Viggiani
Developmental Editor:
Christopher Shortt

Executive Marketing Manager:
Maura Theriault
Executive Production Manager:
Mary Ellen Black
Production Manager:
Larry Main
Production Editor:
Betsy Hough
Production Editor:
Tom Stover

Channel Manager:
Mona Caron
Marketing Coordinator:
Brian McGrath
Cover Design:
Michael Egan
Cover Images Courtesy of:
DaimlerChrysler

Printed in Canada
4 5 6 7 8 9 XXX 05 04 03 02

For more information contact Delmar,
3 Columbia Circle, PO Box 15015,
Albany, NY 12212-5015.

Or find us on the World Wide Web at http://www.delmar.com or
http://www.autoed.com

ISBN: 0-7668-3431-x

NOTICE TO THE READER

Contents

Section 5 Sample Test for Practice

Section 6 Additional Test Questions for Practice

Section 7 Appendices

Preface

This book is just one of a comprehensive series designed to prepare technicians to take and pass every ASE test. Delmar's series covers all of the Automotive tests A1 through A8 as well as Advanced Engine Performance L1 and Parts Specialist P2. The series also covers the five Collision Repair tests and the eight Medium/Heavy Duty truck tests.

Before any book in this series was written, Delmar staff met with and surveyed technicians and shop owners who have taken ASE tests and have used other preparatory materials. We found that they wanted, first and foremost, *lots* of practice tests and questions. Each book in our series contains a sample test and additional practice questions. You will be hard-pressed to find a test prep book with more questions for you to practice with. We have worked hard to ensure that these questions match the ASE style in types of questions, quantities, and level of difficulty.

Technicians also told us that they wanted to understand the ASE test and to have practical information about what they should expect. We have provided that as well, including a history of ASE and a section devoted to helping the technician "Take and Pass Every ASE Test" with case studies, test-taking strategies, and test formats.

Finally, techs wanted refresher information and references. Each of our books includes an overview section that is referenced to the task list. The complete task lists for each test appear in each book for the user's reference. There is also a complete glossary of terms for each booklet.

So whether you're looking for a sample test and a few extra questions to practice with or a complete introduction to ASE testing, with support for preparing thoroughly, this book series is an excellent answer.

We hope you benefit from this book and that you pass every ASE test you take!

Your comments, both positive and negative, are certainly encouraged! Please contact us at:

Automotive Editor
Delmar Publishers
3 Columbia Circle
Box 15015
Albany, NY 12212-5015

1 The History of ASE

History

Originally known as The National Institute for Automotive Service Excellence (NIASE), today's ASE was founded in 1972 as a non-profit, independent entity dedicated to improving the quality of automotive service and repair through the voluntary testing and certification of automotive technicians. Until that time, consumers had no way of distinguishing between competent and incompetent automotive mechanics. In the mid-1960s and early 1970s, efforts were made by several automotive industry affiliated associations to respond to this need. Though the associations were non-profit, many regarded certification test fees merely as a means of raising additional operating capital. Also, some associations, having a vested interest, produced test scores heavily weighted in the favor of its members.

From these efforts a new independent, non-profit association, the National Institute for Automotive Service Excellence (NIASE), was established. In early NIASE tests, Mechanic A, Mechanic B type questions were used. Over the years the trend has not changed, but in mid-1984 the term was changed to Technician A, Technician B to better emphasize sophistication of the skills needed to perform successfully in the modern motor vehicle industry. In certain tests the term used is Estimator A/B, Painter A/B, or Parts Specialist A/B. At about that same time, the logo was changed from "The Gear" to "The Blue Seal," and the organization adopted the acronym ASE for Automotive Service Excellence.

ASE

ASE's mission is to improve the quality of vehicle repair and service in the United States through the testing and certification of automotive repair technicians. Prospective candidates register for and take one or more of ASE's many exams.

Upon passing at least one exam and providing proof of two years of related work experience, the technician becomes ASE certified. A technician who passes a series of exams earns ASE Master Technician status. An automobile technician, for example, must pass eight exams for this recognition.

The exams, conducted twice a year at over seven hundred locations around the country, are administered by American College Testing (ACT). They stress real-world diagnostic and repair problems. Though a good knowledge of theory is helpful to the technician in answering many of the questions, there are no questions specifically on theory. Certification is valid for five years. To retain certification, the technician must be retested to renew his or her certificate.

The automotive consumer benefits because ASE certification is a valuable yardstick by which to measure the knowledge and skills of individual technicians, as well as their commitment to their chosen profession. It is also a tribute to the repair facility employing ASE certified technicians. ASE certified technicians are permitted to wear blue and white ASE shoulder insignia, referred to as the "Blue Seal of Excellence," and carry credentials listing their areas of expertise. Often employers display their technicians' credentials in the customer waiting area. Customers look for facilities that display ASE's Blue Seal of Excellence logo on outdoor signs, in the customer waiting area, in the telephone book (Yellow Pages), and in newspaper advertisements.

To become ASE certified, contact:

National Institute for Automotive Service Excellence
13505 Dulles Technology Drive
Herndon, VA 20171-3421

ASE CERTIFIED

WE SUPPORT VOLUNTARY TECHNICIAN CERTIFICATION THROUGH

National Institute for AUTOMOTIVE SERVICE EXCELLENCE

2 Take and Pass Every ASE Test

ASE Testing

Participating in an Automotive Service Excellence (ASE) voluntary certification program gives you a chance to show your customers that you have the "know-how" needed to work on today's modern vehicles. The ASE certification tests allow you to compare your skills and knowledge to the automotive service industry's standards for each specialty area.

If you are the "average" automotive technician taking this test, you are in your mid-thirties and have not attended school for about fifteen years. That means you probably have not taken a test in many years. Some of you, on the other hand, have attended college or taken postsecondary education courses and may be more familiar with taking tests and with test-taking strategies. There is, however, a difference in the ASE test you are preparing to take and the educational tests you may be accustomed to.

Who Writes the Questions?

The questions on all ASE tests are written by service industry experts familiar with all aspects of the subject area. ASE questions are entirely job-related and designed to test the skills that you need to know on the job.

The questions originate in an ASE "item-writing" workshop where service representatives from domestic and import automobile manufacturers, parts and equipment manufacturers, and vocational educators meet in a workshop setting to share their ideas and translate them into test questions. Each test question written by these experts is reviewed by all of the members of the group.

All of the questions are pretested and quality-checked in a nonscoring section of tests by a national sample of certifying technicians. The questions that meet ASE's high standards of accuracy and quality are then included in the scoring sections of future tests. Those questions that do not pass ASE's stringent test are sent back to the workshop or are discarded. ASE's tests are monitored by an independent proctor and are administered and machine-scored by an independent provider, American College Testing (ACT).

Objective Tests

A test is called an objective test if the same standards and conditions apply to everyone taking the test and there is only one correct answer to each question. Objective tests primarily measure your ability to recall information. A well-designed objective test can also test your ability to understand, analyze, interpret, and apply your knowledge. Objective tests include true-false, multiple choice, fill in the blank, and matching questions. ASE's tests consist exclusively of four-part multiple-choice objective questions.

Before beginning to take an objective test, quickly look over the test to determine the number of questions, but do not try to read through all of the questions. In an ASE test, there are usually between forty and eighty questions, depending on the subject. Read through each question before marking your answer. Answer the questions in the order they appear on the test. Leave the questions blank that you are not sure of and move on to the next question. You can return to those unanswered questions after you have finished the others. They may be easier to answer at a later time after your mind has had additional time to consider them on a subconscious level. In addition, you might find information in other questions that will help you to answer some of them.

Do not be obsessed by the apparent pattern of responses. For example, do not be influenced by a pattern like **d, c, b, a, d, c, b, a** on an ASE test.

There is also a lot of folk wisdom about taking objective tests. For example, there are those who would advise you to avoid response options that use certain words such as *all, none, always, never, must,* and *only,* to name a few. This, they claim, is because nothing in life is exclusive. They would advise you to choose response options that use words that allow for some exception, such as *sometimes, frequently, rarely, often, usually, seldom,* and *normally.* They would also advise you to avoid the first and last option (A and D) because test writers, they feel, are more comfortable if they put the correct answer in the middle (B and C) of the choices. Another recommendation often offered is to select the option that is either shorter or longer than the other three choices because it is more likely to be correct. Some would advise you to never change an answer since your first intuition is usually correct.

Although there may be a grain of truth in this folk wisdom, ASE test writers try to avoid them and so should you. There are just as many **A** answers as there are **B** answers, just as many **D** answers as **C** answers. As a matter of fact, ASE tries to balance the answers at about 25 percent per choice **A, B, C,** and **D.** There is no intention to use "tricky" words, such as outlined above. Put no credence in the opposing words "sometimes" and "never," for example.

Multiple-choice tests are sometimes challenging because there are often several choices that may seem possible, and it may be difficult to decide on the correct choice. The best strategy, in this case, is to first determine the correct answer before looking at the options. If you see the answer you decided on, you should still examine the options to make sure that none seem more correct than yours. If you do not know or are not sure of the answer, read each option very carefully and try to eliminate those options that you know to be wrong. That way, you can often arrive at the correct choice through a process of elimination.

If you have gone through all of the test and you still do not know the answer to some of the questions, then guess. Yes, guess. You then have at least a 25 percent chance of being correct. If you leave the question blank, you have no chance. In ASE tests, there is no penalty for being wrong.

Preparing for the Exam

The main reason we have included so many sample and practice questions in this guide is, simply, to help you learn what you know and what you don't know. We recommend that you work your way through each question in this book. Before doing this, carefully look through Section 3; it contains a description and explanation of the questions you'll find in an ASE exam.

Once you know what the questions will look like, move to the sample test. After you have answered one of the sample questions (Section 5), read the explanation (Section 7) to the answer for that question. If you don't feel you understand the reasoning for the correct answer, go back and read the overview (Section 4) for the task that is related to that question. If you still don't feel you have a solid understanding of the material, identify a good source of information on the topic, such as a textbook, and do some more studying.

After you have completed the sample test, move to the additional questions (Section 6). This time answer the questions as if you were taking an actual test. Once you have answered all of the questions, grade your results using the answer key in Section 7. For every question that you gave a wrong answer to, study the explanations to the answers and/or the overview of the related task areas.

Here are some basic guidelines to follow while preparing for the exam:

- Focus your studies on those areas you are weak in.
- Be honest with yourself while determining if you understand something.
- Study often but in short periods of time.
- Remove yourself from all distractions while studying.
- Keep in mind the goal of studying is not just to pass the exam, the real goal is to learn!

During the Test

Mark your bubble sheet clearly and accurately. One of the biggest problems an adult faces in test-taking, it seems, is in placing an answer in the correct spot on a bubble sheet. Make certain that you mark your answer for, say, question 21, in the space on the bubble sheet designated for the answer for question 21. A correct response in the wrong bubble will probably be wrong. Remember, the answer sheet is machine scored and can only "read" what you have bubbled in. Also, do not bubble in two answers for the same question.

If you finish answering all of the questions on a test ahead of time, go back and review the answers of those questions that you were not sure of. You can often catch careless errors by using the remaining time to review your answers.

At practically every test, some technicians will invariably finish ahead of time and turn their papers in long before the final call. Do not let them distract or intimidate you. Either they knew too little and could not finish the test, or they were very self-confident and thought they knew it all. Perhaps they were trying to impress the proctor or other technicians about how much they know. Often you may hear them later talking about the information they knew all the while but forgot to respond on their answer sheet.

It is not wise to use less than the total amount of time that you are allotted for a test. If there are any doubts, take the time for review. Any product can usually be made better with some additional effort. A test is no exception. It is not necessary to turn in your test paper until you are told to do so.

Your Test Results!

You can gain a better perspective about tests if you know and understand how they are scored. ASE's tests are scored by American College Testing (ACT), a non-partial, non-biased organization having no vested interest in ASE or in the automotive industry. Each question carries the same weight as any other question. For example, if there are fifty questions, each is worth 2 percent of the total score. The passing grade is 70 percent. That means you must correctly answer thirty-five of the fifty questions to pass the test.

The test results can tell you:

- where your knowledge equals or exceeds that needed for competent performance, or
- where you might need more preparation.

The test results *cannot* tell you:

- how you compare with other technicians, or
- how many questions you answered correctly.

Your ASE test score report will show the number of correct answers you got in each of the content areas. These numbers provide information about your performance in each area of the test. However, because there may be a different number of questions in each area of the test, a high percentage of correct answers in an area with few questions may not offset a low percentage in an area with many questions.

It may be noted that one does not "fail" an ASE test. The technician who does not pass is simply told "More Preparation Needed." Though large differences in percentages may indicate problem areas, it is important to consider how many questions were asked in each area. Since each test evaluates all phases of the work involved in a service specialty, you should be prepared in each area. A low score in one area could keep you from passing an entire test.

There is no such thing as average. You cannot determine your overall test score by adding the percentages given for each task area and dividing by the number of areas. It doesn't work that way because there generally are not the same number of questions in each task area. A task area with twenty questions, for example, counts more toward your total score than a task area with ten questions.

Your test report should give you a good picture of your results and a better understanding of your task areas of strength and weakness.

If you fail to pass the test, you may take it again at any time it is scheduled to be administered. You are the only one who will receive your test score. Test scores will not be given over the telephone by ASE nor will they be released to anyone without your written permission.

3 Types of Questions on an ASE Exam

ASE certification tests are often thought of as being tricky. They may seem to be tricky if you do not completely understand what is being asked. The following examples will help you recognize certain types of ASE questions and avoid common errors.

Each test is made up of forty to eighty multiple-choice questions. Multiple-choice questions are an efficient way to test knowledge. To answer them correctly, you must think about each choice as a possibility, and then choose the one that best answers the question. To do this, read each word of the question carefully. Do not assume you know what the question is about until you have finished reading it.

About 10 percent of the questions on an actual ASE exam will use an illustration. These drawings contain the information needed to correctly answer the question. The illustration must be studied carefully before attempting to answer the question. Often, techs look at the possible answers then try to match up the answers with the drawing. Always do the opposite; match the drawing to the answers. When the illustration is showing an electrical schematic or another system in detail, look over the system and try to figure out how the system works before you look at the question and the possible answers.

Multiple-Choice Questions

One type of multiple-choice question has three wrong answers and one correct answer. The wrong answers, however, may be almost correct, so be careful not to jump at the first answer that seems to be correct. If all the answers seem to be correct, choose the answer that is the most correct. If you readily know the answer, this kind of question does not present a problem. If you are unsure of the answer, analyze the question and the answers. For example:

A rocker panel is a structural member of which vehicle construction type?

A. Front-wheel drive

B. Pickup truck

C. Unibody

D. Full-frame

Analysis:

This question asks for a specific answer. By carefully reading the question, you will find that it asks for a construction type that uses the rocker panel as a structural part of the vehicle.

Answer A is wrong. Front-wheel drive is not a vehicle construction type.

Answer B is wrong. A pickup truck is not a type of vehicle construction.

Answer C is correct. Unibody design creates structural integrity by welding parts together, such as the rocker panels, but does not require exterior cosmetic panels installed for full strength.

Answer D is wrong. Full-frame describes a body-over-frame construction type that relies on the frame assembly for structural integrity.

Therefore, the correct answer is C. If the question was read quickly and the words "construction type" were passed over, answer A may have been selected.

EXCEPT Questions

Another type of question used on ASE tests has answers that are all correct except one. The correct answer for this type of question is the answer that is wrong. The word "EXCEPT" will always be in capital letters. You must identify which of the choices is the wrong answer. If you read quickly through the question, you may overlook what the question is asking and answer the question with the first correct statement. This will make your answer wrong. An example of this type of question and the analysis is as follows:

All of the following are tools for the analysis of structural damage EXCEPT:

A. height gauge.

B. tape measure.

C. dial indicator.

D. tram gauge.

Analysis:

The question really requires you to identify the tool that is not used for analyzing structural damage. All tools given in the choices are used for analyzing structural damage except one. This question presents two basic problems for the test-taker who reads through the question too quickly. It may be possible to read over the word "EXCEPT" in the question or not think about which type of damage analysis would use answer C. In either case, the correct answer may not be selected. To correctly answer this question, you should know what tools are used for the analysis of structural damage. If you cannot immediately recognize the incorrect tool, you should be able to identify it by analyzing the other choices.

Answer A is wrong. A height gauge *may* be used to analyze structural damage.

Answer B is wrong. A tape measure may be used to analyze structural damage.

Answer C is correct. A dial indicator may be used as a damage analysis tool for moving parts, such as wheels, wheel hubs, and axle shafts, but would not be used to measure structural damage.

Answer D is wrong. A tram gauge *is* used to measure structural damage.

Technician A, Technician B Questions

The type of question that is most popularly associated with an ASE test is the "Technician A says. . . Technician B says. . . Who is right?" type. In this type of question, you must identify the correct statement or statements. To answer this type of question correctly, you must carefully read each technician's statement and judge it on its own merit to determine if the statement is true.

Typically, this type of question begins with a statement about some analysis or repair procedure. This is followed by two statements about the cause of the problem, proper inspection, identification, or repair choices. You are asked whether the first statement, the second statement, both statements, or neither statement is correct. Analyzing this type of question is a little easier than the other types because there are only two ideas to consider although there are still four choices for an answer.

Technician A, Technician B questions are really double true or false questions. The best way to analyze this kind of question is to consider each technician's statement separately. Ask yourself, is A true or false? Is B true or false? Then select your answer from the four choices. An important point to remember is that an ASE Technician A, Technician B question will never have Technician A and B directly disagreeing with each other. That is why you must evaluate each statement independently. An example of this type of question and the analysis of it follows.

Structural dimensions are being measured. Technician A says comparing measurements from one side to the other is enough to determine the damage. Technician

B says a tram gauge can be used when a tape measure cannot measure in a straight line from point to point. Who is right?

A. A only

B. B only

C. Both A and B

D. Neither A nor B

Analysis:

With some vehicles built asymmetrically, side-to-side measurements are not always equal. The manufacturer's specifications need to be verified with a dimension chart before reaching any conclusions about the structural damage.

Answer A is wrong. Technician A's statement is wrong. A tram gauge would provide a point-to-point measurement when a part, such as a strut tower or air cleaner, interrupts a direct line between the points.

Answer B is correct. Technician B is correct. A tram gauge can be used when a tape measure cannot be used to measure in a straight line from point to point.

Answer C is wrong. Since Technician A is not correct, C cannot be the correct answer.

Answer D is wrong. Since Technician B is correct, D cannot be the correct answer.

Most-Likely Questions

Most-likely questions are somewhat difficult because only one choice is correct while the other three choices are nearly correct. An example of a most-likely-cause question is as follows:

The most likely cause of reduced turbocharger boost pressure may be a:

A. westgate valve stuck closed.

B. westgate valve stuck open.

C. leaking westgate diaphragm.

D. disconnected westgate linkage.

Analysis:

Answer A is wrong. A westgate valve stuck closed increases turbocharger boost pressure.

Answer B is correct. A westgate valve stuck open decreases turbocharger boost pressure.

Answer C is wrong. A leaking westgate valve diaphragm increases turbocharger boost pressure.

Answer D is wrong. A disconnected westgate valve linkage will increase turbocharger boost pressure.

LEAST-Likely Questions

Notice that in most-likely questions there is no capitalization. This is not so with LEAST-likely type questions. For this type of question, look for the choice that would be the least likely cause of the described situation. Read the entire question carefully before choosing your answer. An example is as follows:

What is the LEAST likely cause of a bent pushrod?

A. Excessive engine speed

B. A sticking valve

C. Excessive valve guide clearance

D. A worn rocker arm stud

Analysis:
Answer A is wrong. Excessive engine speed may cause a bent pushrod.
Answer B is wrong. A sticking valve may cause a bent pushrod.
Answer C is correct. Excessive valve clearance will not generally cause a bent pushrod.
Answer D is wrong. A worn rocker arm stud may cause a bent pushrod.

Summary

There are no four-part multiple-choice ASE questions having "none of the above" or "all of the above" choices. ASE does not use other types of questions, such as fill-in-the-blank, completion, true-false, word-matching, or essay. ASE does not require you to draw diagrams or sketches. If a formula or chart is required to answer a question, it is provided for you. There are no ASE questions that require you to use a pocket calculator.

Testing Time Length

An ASE test session is four hours and fifteen minutes. You may attempt from one to a maximum of four tests in one session. It is recommended, however, that no more than a total of 225 questions be attempted at any test session. This will allow for just over one minute for each question.

Visitors are not permitted at any time. If you wish to leave the test room, for any reason, you must first ask permission. If you finish your test early and wish to leave, you are permitted to do so only during specified dismissal periods.

You should monitor your progress and set an arbitrary limit to how much time you will need for each question. This should be based on the number of questions you are attempting. It is suggested that you wear a watch because some facilities may not have a clock visible to all areas of the room.

4 An Overview of the System

Engine Performance (Test A8)

The following section includes the task areas and task lists for this test and a written overview of the topics covered in the test.

The task list describes the actual work you should be able to do as a technician that you will be tested on by the ASE. This is your key to the test and you should review this section carefully. We have based our sample test and additional questions upon these tasks, and the overview section will also support your understanding of the task list. ASE advises that the questions on the test may not equal the number of tasks listed; the task lists tell you what ASE expects you to know how to do and be ready to be tested upon.

At the end of each question in the Sample Test and Additional Test Questions sections, a letter and number will be used as a reference back to this section for additional study. Note the following example: **C.12.**

Task List

C. Fuel, Air Induction, and Exhaust System Diagnosis and Repair (14 Questions)

Task C.12 **Inspect, clean, or replace throttle body mounting plates, air induction system, intake manifold, and gaskets.**

Example:

1. While discussing engine performance diagnosis, Technician A says a vacuum leak decreases engine performance. Technician B says propane is the best method of locating vacuum leaks. Who is right?
   A. A only
   B. B only
   C. Both A and B
   D. Neither A nor B (C.12)

Analysis:

Question #1
Answer A is wrong.
Answer B is wrong.
Answer C is correct.
Answer D is wrong.

Task List and Overview

A. General Engine Diagnosis (10 Questions)

Task A.1 **Verify driver's complaint, perform visual inspection, and/or road test vehicle; determine needed action.**

When a customer brings in their vehicle for service, they have a concern. The first step in the diagnostic procedure is to collect information about the vehicle and the customer's concern. This can be accomplished by talking to the customer or reviewing the work order with the service advisor. A road test should be performed to confirm the problem. Then a visual inspection of the vehicle should be made to determine if there is an obvious reason for the problem.

Task A.2 **Research applicable vehicle information, such as engine management system operation, vehicle service history, service precautions, and technical service bulletins.**

An important part of diagnosis is knowing the system you are working on. This applies to engine and powertrain management systems as well as other parts of the vehicle. Often this information includes special service and diagnostic procedures or system changes that are developed after the printing of the factory service manual. There are many sources for this information; they can be electronic or printed.

Task A.3 **Diagnose the cause of unusual engine noise and/or vibration problems; determine needed action.**

Engine defects such as damaged pistons, worn rings, loose piston pins, worn crankshaft bearings, worn camshaft lobes, and loose or worn valve train components usually produce their own identifiable noises or vibrations. Identifying when the noises and vibrations occur can be helpful in determining the faulty component. A stethoscope, for example, would be useful in determining the location of a noise.

Task A.4 **Diagnose the cause of unusual exhaust color, odor, and sound; determine needed action.**

If the engine is operating normally, the exhaust should be colorless. A small swirl of white vapor from the tailpipe is normal in cold weather. This is vapor moisture in the exhaust and is a byproduct of combustion. If the exhaust is blue, small amounts of oil are entering the combustion chamber. If the exhaust is black, the air-fuel mixture is too rich. If the exhaust is gray, coolant may be leaking into the combustion chambers.

Task A.5 **Perform engine manifold vacuum or pressure tests; determine needed action.**

When a vacuum gauge is connected to the intake manifold, the gauge should provide a steady reading between 17 and 22 in. Hg (44.8 and 27.6 kPa absolute) with the engine idling. A low steady reading indicates late ignition timing. Burned or leaking valves cause a vacuum gauge fluctuation between 12 and 18 in. Hg (and 41.4 kPa absolute). When the engine is accelerated and held at a steady higher revolutions per minute (rpm) and the vacuum slowly drops to a very low reading, the exhaust system is restricted.

When performing vacuum tests, the technician should keep in mind the effect of the valve train on the production of vacuum. If the valve timing is not correct the engine will not perform as designed and lower vacuum readings will result. Additionally, incorrect valve adjustment can compromise engine efficiency. Both valve adjustment and valve timing should be checked in cases of poor vacuum production.

Task A.6 Perform cylinder power balance test; determine needed action.

A cylinder power balance test is performed to ensure that all cylinders are contributing equally. When individual cylinders are disabled, a noticeable revolutions per minute (rpm) drop is measured. By comparing the amount of rpm drop between cylinders, it can be determined which cylinder has the problem.

Task A.7 Perform cylinder cranking compression test; determine needed action.

The compression test checks the sealing quality of the combustion chamber. If the compression is lower than specified in one or more cylinders, then the valves and rings are suspect. A wet compression test is useful in deciding whether rings or valves are the problem. If compression comes up during the wet compression test, then rings are the most likely problem. Conversely, if compression does not come up, then leaking valves are the likely problem. When performing a compression test the engine should be at operating temperature and the throttle should be held open in order to get a more accurate reading.

If it is determined that the valves are the cause of compression loss, the valve clearance should be checked and adjusted before continuing with repair. In some cases valve timing could cause compression loss. If this is the case the sound of the engine while cranking is distinctive and will lead the technician to investigate valve timing.

Task A.8 Perform cylinder leakage test; determine needed action.

The cylinder leakage test (leak-down test) can be used to further pinpoint the problem. A regulated amount of air is introduced into the cylinder and the gauge on the tester will indicate the percentage of that pressure which is leaking. A gauge reading of 0 percent indicates no leakage while a reading of 100 percent indicates that cylinder does not hold compression. If two adjacent cylinders have excess leakage, a head gasket problem is likely. While testing a cylinder with high leakage, you should try to find where the leakage is going. For example if air is heard escaping from the exhaust, then a leaking exhaust valve is indicated. Air coming from the intake would indicate a leaking intake valve. Leakage could show up in the radiator, indicating a problem with a head gasket or a cracked cylinder head. Air coming from the PCV valve would indicate leakage past the rings.

Task A.9 Diagnose engine mechanical, electrical, electronic, fuel, and ignition problems with an oscilloscope and/or engine analyzer; determine needed action.

Today's engine analyzers have capabilities ranging from starting system diagnosis to multi-trace wave form analysis. The multi-trace function allows the technician to observe a signal in the context of the relationship between time and voltage. A single trace can easily demonstrate a "lazy" oxygen sensor response, for example. Multiple traces allow comparison between several signals, such as secondary ignition traces.

Task A.10 Prepare and inspect vehicle and analyzer for exhaust gas analysis; obtain exhaust gas readings.

Many states have emission inspection programs that require vehicle owners to maintain their vehicles to certain standards. An emissions analyzer measures tailpipe emissions. Emission analyzers require a warm-up period and certain calibration intervals. Some items that may be checked with an emissions analyzer include: air-fuel mixture, cylinder misfire, catalytic converter defects, and leaking head gaskets. A four-gas emissions analyzer is capable of measuring hydrocarbons, carbon monoxide, oxygen, and carbon dioxide.

Task A.11 Verify correct valve adjustment on engines with mechanical or hydraulic lifters.

Valve lifters are either mechanical (solid) or hydraulic. Solid lifters provide for a rigid connection between the camshaft and the valves. Hydraulic valve lifters provide for the same connection but use oil to absorb the shock that results from the movement of the valve train.

Hydraulic lifters are designed to automatically compensate for the effects of engine temperature. Changes in temperature cause the valve train components to expand and contract. Solid lifters require a clearance between the parts of the valve train. This clearance allows for the expansion of the components as the engine gets hot. Periodic adjustment of this clearance must be made. Excessive clearance might cause a clicking noise. This clicking noise is also an indication of the hammering of the valve train parts against one another, which will result in reduced camshaft and lifter life.

Valve lash on some engines is adjusted with an adjusting nut on the valve tip end of the rocker arm. The clearance is checked by inserting a feeler gauge between the valve tip and the adjusting nut. Some OHC engines have an adjustment disc or shim between the cam lobe surface and the lifter or follower. To adjust valve lash, a special tool and a magnet must be used.

Task A.12 Verify correct camshaft timing; determine needed action.

The camshaft and crankshaft must always remain in the same relative position to each other. They must also be in the proper initial relation to each other. This initial position between the shafts is designated by timing marks. To obtain the correct initial relationship of the components, the corresponding marks are aligned during engine assembly. Verification of this relationship is done by rotating the crankshaft to TDC on cylinder #1 and checking the alignment of the timing marks on both shafts.

Task A.13 Verify proper engine operating temperature, check coolant level and condition, perform cooling system pressure test; determine needed repairs.

The cooling system must operate, be inspected, and be serviced as a system. Replacing one damaged part while leaving others dirty or clogged will not increase system efficiency. Service the entire system to ensure good results. Service involves both a visual inspection of the parts and connections and pressure testing. Pressure testing is used to detect internal or external leaks. Pressure testing can also be used to check the condition of the radiator cap. This type of testing involves applying a pressure to the system or cap. If the system is able to hold the pressure, there are no leaks in the system. If the pressure drops, there is an internal or external leak.

The vehicle's temperature gauge, a shop temperature gauge, or a hand-held pyrometer can verify engine-operating temperature. The condition and level of the coolant should be checked as part of a preventative maintenance program. The level of the coolant should be at the level specified by the manufacturer. The coolant should be checked for the presence of engine oil or other contaminants.

Task A.14 Inspect, test, and replace mechanical/electrical fans, fan clutch, fan shroud/ducting, and fan control devices.

Mechanical fans can be checked by spinning the fan by hand. A noticeable wobble or any blade that is not in the same plane as the rest indicates that the fan needs to be replaced. One of the simplest checks of a fan clutch is to look for signs of fluid loss. Oily streaks radiating outward from the hub shaft mean fluid has leaked out past the bearing seal. Most fan clutches offer a slight amount of resistance if turned by hand when the engine is cold. They offer drag when the engine is hot. If the fan freewheels easily when it is hot or cold, replace the clutch.

Electric cooling fans are mounted to the radiator shroud and are not mechanically connected to the engine. An electric motor-driven fan is controlled by either, or both, an engine coolant temperature switch or sensor and an air conditioning switch. The controls for the electric cooling fan can be easily identified by referring to the wiring schematic for the vehicle.

B. Ignition System Diagnosis and Repair (13 Questions)

Task B.1 Diagnose no-starting, hard starting, engine misfire, poor driveability, spark knock, power loss, poor mileage, and emissions problems on vehicles with distributor and distributorless ignition systems; determine needed repairs.

When performing no-start ignition diagnostics, the first step is to check for available spark at a plug wire. Then, connect a test light between the negative side of the coil and ground. If the test light flutters on and off when the key is placed in the CRANK position, the pickup coil signal and module are functioning properly and the secondary ignition system should be tested. If the test light does not flutter, test the pickup coil with an ohmmeter. If the pickup coil is satisfactory, the module is defective.

Many modern vehicles have a distributorless ignition system (DIS). These systems use a variety of sensors to achieve proper spark timing. These systems generally have one ignition coil per cylinder. In some cases there is one coil for two cylinders. This system fires the coil twice per revolution which will fire both spark plugs twice, once on compression and once on exhaust (so-called waste spark).

Task B.2 Check for possible ignition system related diagnostic trouble codes (DTCs).

DTCs can be retrieved from the PCM on nearly all vehicles. The DTCs are displayed in the instrument panel or on a scan tool. The latter is the most common. The displayed trouble codes are interpreted in the service manual. This interpretation identifies the area of the vehicle that has triggered the DTC. This area is not always where the problem exists. Further diagnostics is required to properly identify the exact problem after retrieving a DTC.

Task B.3 Inspect, test, repair, or replace ignition primary circuit wiring and components.

Many ignition module testers are available from vehicle and test equipment manufacturers. These testers check the module's capability of switching the primary ignition circuit on and off. On some testers, a green light is illuminated if the module is satisfactory, and the light remains off when the module is defective. Always follow the manufacturer's recommended procedure. If the module tests satisfactory, then the technician should perform circuit tests to confirm that the wiring of the circuit is serviceable and the proper signals are going to the correct destinations.

Task B.4 Inspect, test, and service distributor.

The distributor should be checked to confirm that it is functioning correctly mechanically. Bushings, centrifugal advance, and body condition should be evaluated and repaired as required. Distributor testers are available that can test the distributor dynamically. This is particularly useful in checking the effect of shaft play on the performance of the distributor.

Before installing and timing the distributor, ensure that the engine is at top dead center (TDC) on the compression stroke of the specified cylinder. In most cases, this is cylinder no. 1, but some manufactures specify a different cylinder for particular engines. Position the distributor rotor so it will point toward the specified cap terminal when the distributor cap is installed.

Task B.5 Inspect, test, service, repair, or replace ignition system secondary circuit wiring and components.

Check spark plug wire's physical condition. Replace if the insulation is damaged or oil/coolant soaked. Check resistance of wires while still attached to the distributor cap. This will test both the wire and the terminal in the cap. Consult the workshop manual for the resistance specification, usually expressed as ohms per foot of wire. Replace as necessary.

A scope is very useful for determining which cylinder is affected. Set the scope up to show the secondary ignition as a parade trace; the cylinder with the problem is easily identified.

Check the distributor cap and rotor for damage and replace as necessary. On some vehicles there is a resistor in the distributor rotor; check the workshop manual for the specification.

Task B.6 Inspect, test, and replace ignition coil(s).

The ignition coil should be inspected for cracks or any evidence of leakage in the coil tower. The coil container should be checked for oil leaks. If oil is leaking from the coil, air space is present, allowing condensation to form internally. Condensation in an ignition coil causes voltage leaks and engine misfire. When testing the coil with an ohmmeter, most primary windings have a resistance of 0.5 ohms to 2 ohms and secondary windings have a resistance of 8,000 ohms to 20,000 ohms. The maximum coil output can be tested with an engine analyzer. Always refer to the manufacturer's specifications.

Task B.7 Check and adjust ignition system timing and timing advance/retard.

Ignition timing specifications and instructions are included on the under-hood emissions label. A timing light is connected to the #1 cylinder spark plug wire and to the vehicle battery. The vehicle must be at the specified revolutions per minute (rpm) when the timing light is aimed at the timing indicators. Observe the timing marks. If the timing marks are not at the specified location, rotate the distributor until the mark is at the specified location and tighten down the distributor.

Task B.8 Inspect, test, and replace ignition system pickup sensor or triggering devices.

Connect an ohmmeter to the pickup coil terminals and pull on the pickup leads. Erratic meter readings indicate an open in the pickup leads. Most pickup coils have 150 ohms to 900 ohms of resistance, but always refer to the manufacturer's specification. A reading above 900 ohms indicates an open pickup coil. A reading below 150 ohms indicates a shorted pickup coil. Connect one ohmmeter lead to one of the pickup leads and the other to ground to test the pickup for a grounded condition. An infinite reading means the pickup is not grounded.

Task B.9 Inspect, test, and replace ignition control module.

The ignition module removal and replacement procedure varies, depending on the ignition system. Always follow the manufacturer's recommended replacement procedure. Some ignition modules require the use of dielectric silicone grease for heat dissipation through the mounting surface. Wipe the mounting surface clean and place a light coating of silicone on the ignition module. If silicone is not used, heat will not dissipate properly, and damage to the module may occur.

C. Fuel, Air Induction, and Exhaust System Diagnosis and Repair (14 Questions)

Task C.1 Diagnose fuel system related problems—hot or cold no-starting, hard starting, poor driveability, incorrect idle speed, poor idle, flooding, hesitation, surging, engine misfire, power loss, stalling, poor mileage, dieseling, and emission problems on vehicles with injection-type or carburetor-type fuel systems; determine needed action.

In an electronic fuel injection (EFI) system, the control unit must have information from a variety of sensors. These sensors provide information that the control module must have to control the various systems involved. Failure of a sensor will force the control unit to substitute fixed values for the missing sensor data. This will result in compromised performance and, in some cases failure to start. Following is an example of one sensor system and its effect on engine performance.

In an electronic fuel injected (EFI) system, the computer must know the amount of air entering the engine so it can supply the stoichiometric air:fuel ratio. In EFI systems with a manifold absolute pressure (MAP) sensor, the powertrain control module (PCM) calculates the amount of air entering the engine by comparing the MAP sensor to revolutions per minute (rpm) input signals. The PCM supplies a five-volt reference signal and the MAP sensor modifies this signal which returns to the PCM. By monitoring the signal line, the technician can determine if the MAP sensor is functioning normally. A defective MAP sensor may cause a rich or lean air:fuel ratio, excessive fuel consumption, and engine surging.

Task C.2 Check for possible fuel or induction system related diagnostic trouble codes (DTCs).

DTCs can be retrieved from the PCM on nearly all vehicles. The DTCs are displayed in the instrument panel or on a scan tool. The latter is the most common. The displayed trouble codes are interpreted in the service manual. This interpretation identifies the area of the vehicle that has triggered the DTC. This area is not always where the problem exists. Further diagnostics is required to properly identify the exact problem after retrieving a DTC.

Task C.3 Perform fuel system pressure and volume tests; determine needed action.

In order to deliver fuel to the engine properly, fuel must be available to the system at the correct pressure and in adequate volume. It is possible to have the correct pressure with little or no volume. Fuel volume is the amount of fuel delivered over a specified time. This is specified by the manufacturer. Generally about 1 pint (0.5 liter) per 30 seconds is acceptable. Because this test involves discharging fuel into an open container, extreme caution should be exercised to avoid injury or fire.

A mechanical fuel pump can be tested by installing a pressure/vacuum gauge onto the inlet side of the pump to test the pump diaphragm and valve.

Low fuel pressure may cause lack of power, acceleration stumbles, engine surging, and limited top speed, whereas high fuel pressure results in excessive fuel consumption, rough idle, engine stalling, and excessive sulfur smell. On multiport fuel injected (MFI) and sequential fuel injected (SFI) systems, connect the fuel pressure gauge to the Schrader valve test port on the fuel rail. On some throttle body injected (TBI) systems, the gauge is installed between the inlet line and fitting. On other TBI systems, the gauge is installed at the fuel filter inlet. If the pressure is lower than the manufacturer's specifications, check the fuel lines and hoses, fuel pump, filter, and regulator. If pressure is higher than the specifications, check the fuel return line and pressure regulator.

Task C.4 Inspect fuel tank, tank filter, and gas cap; inspect and replace fuel lines, fittings, and hoses; check fuel for contaminants and quality.

The fuel tank should be inspected for leaks; road damage; corrosion; rust; loose, damaged, or defective seams; loose mounting bolts; and damaged mounting straps. Leaks in the fuel tank, lines, or filter may cause gasoline odor in and around the vehicle, especially during low speed driving and idling. In most cases, the fuel tank must be removed for service.

Obtain a sample of the fuel and examine for dirt or other contaminants. Use a commercially available alcohol tester to determine the percentage of alcohol present in the fuel. Generally fuel injection systems will tolerate only a limited amount of alcohol in the fuel. Consult workshop manual for details.

Task C.5 Inspect, test, and replace mechanical and electrical fuel pumps and pump control systems; inspect, service, and replace fuel filters.

When testing mechanical and/or electric fuel pumps the pressure and volume tests apply. Fuel gauge pressure and volume over a given period of time, compared to the manufacturer's specifications, give the technician an accurate diagnostic route to follow. Low pressure or volume could be the result of a restricted fuel filter or fuel line. High pressure or volume could be the result of an inoperative pressure regulator or return line.

Task C.6 Inspect, test, and repair or replace fuel pressure regulation system and components of injection-type fuel systems.

During a pressure test, the fuel pressure regulator is tested by removing the vacuum line during idle operation. A pressure increase should be noted of approximately 10 PSI (69 kPa). If the fuel pressure is low, restrict the return line; if pressure then increases, the regulator is stuck.

Task C.7 Inspect, test, adjust and repair or replace cold enrichment, acceleration enrichment, and deceleration fuel reduction or shut-off components.

Late-model carburetors used on engines before they were fitted with fuel injection used many different systems to help control their operation. Cold enrichment, acceleration enrichment, deceleration fuel reduction, and fuel shut-off systems were some of these systems.

Carburetors have always been equipped with some sort of cold enrichment system. This is what a choke was. Later model carburetors used electrically controlled solenoids that allowed additional fuel into the system when the PCM determined it was best to do so. Likewise, carburetors had acceleration pump systems to enrichen the mixture during acceleration. Some carburetors were equipped with systems that allowed for additional fuel when the throttle plates were fully opened.

Many fuel injection systems also have separate units that prevent, increase, or reduce the amount of fuel being sprayed out of an injector. Most new systems, however, have these controls as part of the PCM's control of the entire system. Enrichment occurs by increasing the on-time of an injector and fuel is reduced by decreasing the on-time.

Task C.8 Remove, clean, and replace throttle body; adjust related linkages.

After many miles of operation, an accumulation of gum and carbon deposits may occur around the throttle area on throttle body injected (TBI), sequential fuel injected (SFI), and multiport fuel injected (MFI) systems. This may cause rough idle operation. Throttle body cleaner may be used to spray around the throttle area without removing and disassembling the throttle body. If this cleaning method does not remove the deposits, the throttle body will have to be removed according to the manufacturer's recommendations, disassembled, and placed in an approved cleaning solution. The throttle position sensor (TPS), idle air control (IAC), fuel injector, pressure regulator, and seals must be removed prior to placing the throttle body in the cleaning solution. Since MFI and SFI systems do not have the pressure regulator or fuel injector in the throttle body, these items need not be removed.

Task C.9

Inspect, test, clean, and replace fuel injectors.

Tool manufacturers market a variety of fuel injector cleaning equipment. A solution of fuel injector cleaner is mixed with unleaded gasoline. Shop air pressure provides system operating pressure. The vehicle's fuel pump must be disabled to prevent fuel from being forced into the fuel rail. The fuel return line should be plugged to keep the fuel injector cleaner solution from entering the fuel tank. After the fuel injectors have been cleaned, the adaptive memory will need to be reset.

If the injectors have been removed for cleaning, the spray pattern should be checked. An even cone shaped pattern without thready or dripping discharge should be present. If the proper spray pattern cannot be achieved the injector should be replaced.

Task C.10

Inspect, service, and repair or replace air filtration system components.

If a vehicle is operated continually in dusty conditions, air filter replacement may be necessary at more frequent intervals. A damaged air filter can cause increased wear on cylinder walls, pistons, and piston rings. If the air filter is restricted with dirt, it restricts the flow of air into the intake manifold, and this increases fuel consumption.

Task C.11

Inspect, clean, or replace throttle body mounting plates, air induction system, intake manifold, and gaskets.

Vacuum leaks can be located using a propane cylinder equipped with a metering valve and hose. When propane is discharged near a leak, the engine will run smoother and the idle will increase. The air inlet tubing on turbocharged engines should be checked for looseness or leaks. A leak in the tubing between the turbo and the intake manifold can reduce the boost pressure available to the engine.

Task C.12

Check/adjust idle speed and fuel mixture.

Before idle mixture adjustment is performed, idle speed and base ignition timing must be set to specifications. All vacuum hose connections and underhood wiring connectors should be checked. When the idle mixture is adjusted using the vehicle manufacturer's recommended procedure, the vehicle should meet emission standards, assuming the engine compression and ignition system are satisfactory. There are two methods of adjusting the idle mixture, CO and lean drop.

The lean drop method is used primarily on older vehicles. This method involves adjusting the mixture leaner until the engine rpm drops and then richening until the idle becomes smooth out.

The CO method involves adjusting the idle until the specified CO reading is obtained. There are more specific details to both these procedures and the workshop manuals should be consulted before repair is attempted.

Task C.13

Remove, clean, inspect/test, and repair or replace vacuum and electrical components and connections of fuel systems.

Nylon fuel pipes should be inspected for leaks, nicks, scratches, cuts, kinks, melting, and loose fittings. If these fuel pipes are kinked or damaged in any way, they must be replaced. Nylon fuel pipes provide a certain amount of flexibility and can be formed around gradual curves under the vehicle. Do not force a nylon fuel pipe into a sharp bend because this action may kink the pipe and restrict the flow of fuel.

Task C.14

Inspect, service, and replace exhaust manifold, exhaust pipes, mufflers, resonators, catalytic converters, tail pipes, and heat shields.

Remove the exhaust pipe bolts at the manifold flange, and disconnect any other components in the manifold, such as the O_2 sensor. Remove the bolts retaining the manifold to the cylinder head, and lift the manifold from the engine compartment. Remove the manifold heat shield. Thoroughly clean the manifold and cylinder head mating surfaces. Measure the exhaust manifold surface for warping with a straightedge and feeler gauge in three locations on the manifold surface. Examine manifold carefully for any cracks or broken flanges.

Follow the exhaust system from manifold to tail pipe end. Ensure that all hangers are present and installed correctly. The exhaust system is designed to be suspended from these hangers; loosen joints and realign if any of the hangers are in tension. Examine all pipes, mufflers, and resonators to ensure that they are securely connected and gas tight.

Task C.15

Perform exhaust system backpressure tests; determine needed action.

Excessive exhaust backpressure may be caused by a restricted exhaust pipe, catalytic converter, or muffler. If the exhaust backpressure is excessive, engine power and maximum vehicle speed are reduced, but the engine does not misfire. Connect a vacuum gauge to the intake manifold to check for a restricted exhaust system. With the engine idling, manifold vacuum should be 16 to 21 in. Hg (48.3 to 31 kPa absolute). When the engine is accelerated to 2,000 rpm and held for three minutes, the vacuum should drop momentarily and then recover. If after the three minutes the vacuum drops below minimum specification, the exhaust system is restricted.

Task C.16

Inspect, test, clean, and repair or replace turbocharger/supercharger and system components.

Both a turbocharger and supercharger increase the amount of air delivered to an engine's cylinders by increasing the amount of pressure the air is delivered at. A turbocharger is driven by the movement and heat of the exhaust leaving the engine. Intake air pressure is increased by a compressor in the turbocharger unit. The faster the compressor turns, the more the air is boosted. The speed of the compressor is determined by the load and speed on the engine. To control the boost and therefore prevent over-boost, turbochargers are equipped with a waste gate that controls the amount of exhaust gas at the turbocharger.

A supercharger is driven by the engine's crankshaft via a drive belt. The speed of the compressor or supercharger is directly related to the speed of the engine. The high pressure from a supercharger is typically controlled via an electromagnetic clutch at the belt drive. When pressure gets too high, the clutch is disengaged. Often the output pressure is controlled by pulsing the drive clutch.

The pressure boost for either system can be measured with a pressure gauge connected to the intake manifold. During a road test, the pressure can be observed during a variety of speed and load conditions. Recording the condition and the resulting pressure can lead to a thorough evaluation of the turbocharger or supercharger system.

Both a supercharger and turbocharger are non-serviceable items. If there is a problem with either unit, it is replaced. Only the control circuits of these systems can be serviced.

D. Emissions Control Systems Diagnosis and Repair (9 Questions)

1. Positive Crankcase Ventilation (PCV) (1 Question)

Task D.1.1

Diagnose emissions or driveability problems caused by PCV system.

If the positive crankcase ventilation (PCV) valve is stuck in the open position, excessive airflow through the valve causes a lean air:fuel ratio and possible rough idle operation or engine stalling. When the PCV valve or hose is restricted, excessive crankcase pressure forces blowby gases through the clean air hose and filter into the air cleaner. Worn rings or cylinders cause excessive blowby gases and increased crankcase pressure, which forces blowby gases through the clean air hose and filter into the air cleaner.

Task D.1.2 Inspect, service, and replace positive crankcase ventilation (PCV) filter/breather cap, valve, tubes, orifices, and hoses.

A thorough examination of the positive crankcase ventilation (PVC) system is relatively easy. After performing the recommended diagnostics, visually inspect the cap, tubes, and hoses for kinks, cuts, or other damage. Disassemble the PVC system to isolate the cause of the restriction. Shake the PCV valve next to your ear and listen for the tapered valve rattling inside the housing. If no rattle is heard, replace the PCV valve.

Positive crankcase ventilation (PCV) diagnostic recommendations differ from manufacturer to manufacturer. Some recommend removing the PCV valve and hose from the rocker cover. Connect a length of hose to the inlet side of the PCV valve, and blow air through the valve with your mouth while holding your finger near the valve outlet. Air should pass freely through the valve. If not, replace the valve. Connect a length of hose to the outlet side of the PCV valve and try to blow back through the valve. If air passes easily through the valve, it should be replaced. Other manufacturers recommend disconnecting one end of the PCV valve and placing a finger over it with the engine idling. When there is no vacuum at the PCV valve, part of the system is restricted.

2. Exhaust Gas Recirculation (EGR) (3 Questions)

Task D.2.1 Test and diagnose emissions or driveability problems caused by the EGR system.

If the exhaust gas recirculation (EGR) valve remains open at idle and low engine speed, the idle operation is rough and surging occurs at low speed. When this problem is present, the engine may hesitate on low-speed acceleration or stall after deceleration or after a cold start. If the EGR valve does not open, engine detonation can occur and emission levels increase.

Task D.2.2 Check for possible exhaust gas recirculation (EGR) related diagnostic trouble codes (DTCs).

DTCs can be retrieved from the PCM on nearly all vehicles. The DTCs are displayed in the instrument panel or on a scan tool. The latter is the most common. The displayed trouble codes are interpreted in the service manual. This interpretation identifies the area of the vehicle that has triggered the DTC. This area is not always where the problem exists. Further diagnostics is required to properly identify the exact problem after retrieving a DTC.

Task D.2.3 Inspect, test, service, and replace components of the EGR system, including EGR tubing, exhaust passages, vacuum/pressure controls, filters, hoses, electrical/electronic sensors, controls, solenoids, and wiring of exhaust gas recirculation (EGR) systems.

The first step in diagnosing any EGR system is to check all of the system's vacuum and electrical connectors. In many systems, the PCM uses inputs from various sensors to operate the EGR valve. Improper EGR operation may be caused by a defect in one or more of the sensors. DTCs should be retrieved and the cause corrected before any further diagnostics is completed.

With the engine at normal operating temperature and at idle speed, disconnect the vacuum hose at the valve and supply 18 in. Hg (41.4 kPa absolute) of vacuum to the EGR valve and observe the valve's diaphragm and the engine's operation. The EGR valve diaphragm should open and the engine's idle should become erratic. If the valve doesn't hold the applied vacuum or if the valve doesn't open, the valve needs to be replaced. If the valve opens but does not affect engine idle, remove the valve and clean it and related passages as needed.

Often EGR problems are caused by faulty EGR controls, such as the EGR vacuum regulator (EVR). This regulator can be checked with a scan tool or an ohmmeter. Connect the meter across the terminals of the EVR. An infinite reading indicates there is an open inside the EVR whereas as a low resistance reading means the EVR's coil is shorted internally. The coil should also be checked for shorts to ground. To do this, connect the meter at one of the EVR terminals and the other to the case. The reading should be infinite. If there is any measured resistance, the unit is shorted.

Other EGR control components can also be checked on the scan tool or with a DMM. Refer to the appropriate service manual for the exact procedures and the desired specifications.

3. Exhaust Gas Treatment (2 Questions)

Task D.3.1 Test and diagnose emissions or driveability problems caused by the secondary air injection or catalytic converter systems.

Catalytic converters have been in use since 1975. They are placed in the exhaust system shortly after the engine and are designed as "afterburners" to clean the exhaust of excess pollutants. There are three basic types of converters:

- Two-way converters are used primarily on pre-1980 vehicles and control unburned hydrocarbons (HC) and carbon monoxide (CO).
- Three-way converters (TWC) control HC, CO, and oxides of nitrogen (NO_X). These converters were used on 1980-1985 vehicles without computerized engine controls or air injection.
- Three-way plus oxidation converters are used on 1980 and later vehicles with computerized engine controls and air injection.

The most common reasons for converter failure are overheating and fuel poisoning (using leaded gasoline). An engine misfire can allow unburned fuel to enter the converter. This can cause excessive heat and converter failure.

Task D.3.2 Check for possible exhaust gas recirculation (EGR) related diagnostic trouble codes (DTCs).

Check all hoses and pipes in the system for looseness and rusted or burned conditions. Inspect the check valves and replace them if they show signs of leakage. With the engine idling, listen for noises from the pump (if equipped). Check the air pump drive belt; adjust it if loose or replace it if worn or damaged.

On a pump system, follow the air pump outlet hose from the pump to its first connection and disconnect it. With the engine idling, air should flow from the hose. If it does not, check the drive belt and the pump itself. Raise engine speed to 1500 rpm; airflow should increase. If not, check for a slipping drive belt. Remove clean air hose from the air cleaner and start the engine. With the engine idling, there should be steady audible pulses heard at the end of the hose. If these pulses are erratic, check for cylinder misfiring. If the cylinders are not misfiring, check for sticking one-way valves or restricted exhaust inlet air tubes in the exhaust manifold.

Task D.3.3 Inspect, test, service, and replace mechanical components and electrical/electronically operated components and circuits of secondary air injection systems.

For a short period of time after the engine is started, you should be able to hear air being exhausted from the secondary air injection bypass (AIRB) valve. If this air is not being exhausted, remove the vacuum hose from the AIRB valve. If air is now being exhausted from the AIRB valve, check the AIRB solenoid and connecting circuits. When air is still not being exhausted from the AIRB valve, check the air supply from the pump to the valve. If air supply is available, replace the AIRB valve. If air supply is not available, replace the pump.

Task D.3.4

Inspect and test the catalytic converter(s).

If the catalytic converter rattles when tapped with a soft hammer, the internal components are loose and the converter should be replaced. When a catalytic converter is restricted, a significant loss of power and limited top speed will be noticed.

4. Evaporative Emissions Controls (3 Questions)

Task D.4.1

Diagnose emissions or driveability problems caused by the evaporative emissions control system.

If the EVAP system is purging vapors from the charcoal canister when the engine is idling, rough engine operation will occur. Cracked hoses or a canister saturated with gasoline may allow gasoline vapors to escape to the atmosphere, resulting in gasoline odor in and around the vehicle.

Later model vehicles have an advanced EVAP system. This system uses a fuel tank pressure sensor to detect leaks in the EVAP system. Any leaks will register a DTC and light the malfunction indicator lamp (MIL).

Task D.4.2

Check for possible evaporative emission related diagnostic trouble codes (DTCs).

DTCs can be retrieved from the PCM on nearly all vehicles. The DTCs are displayed in the instrument panel or on a scan tool. The latter is the most common. The displayed trouble codes are interpreted in the service manual. This interpretation identifies the area of the vehicle that has triggered the DTC. This area is not always where the problem exists. Further diagnostics is required to properly identify the exact problem after retrieving a DTC.

With the tank pressure control valve removed, use a vacuum/pressure pump to apply pressure to the tank side of the valve. Some restriction to airflow should be noticed until the air pressure opens the valve. Connect a second vacuum pump to the vacuum fitting on the valve and apply 10 in. Hg (69 kPa absolute) to the valve. Then try to apply pressure through the valve from the tank side. Under this condition, there should be no restriction to airflow. If the tank pressure control valve does not operate properly, replace the valve. If the fuel tank has a pressure valve and a vacuum valve in the filler cap, check these valves for dirt contamination and damage. The cap may be washed in clean solvent. When the valves are sticking or damaged, replace the cap. If the charcoal canister has a replaceable filter, check the filter for dirt contamination. Replace the filter as required.

E. Computerized Engine Controls Diagnosis and Repair (17 Questions)

Task E.1

Retrieve and record stored DTCs.

On some vehicles, the diagnostic trouble codes (DTCs) are obtained by connecting a jumper wire to the proper terminals in the data link connector and then turning on the ignition switch. On other vehicles, the ignition switch is cycled on and off three times in a five-second period to signal the powertrain control module (PCM) to enter the diagnostic mode and provide DTCs. On some vehicles, the DTCs may be read by counting the flashes of the malfunction indicator lamp (MIL). Most technicians use a scan tool for DTC retrieval.

Task E.2 Diagnose the causes of emissions or driveability problems resulting from failure of computerized engine controls with stored DTCs.

Once a fault has been detected by the powertrain control module (PCM), it stores a diagnostic trouble code (DTC) in memory, and if the fault affects exhaust emissions, it will light the malfunction indicator lamp (MIL). The technician then retrieves the stored DTC and accesses a diagnostic flow chart. The diagnostic flow chart leads the technician through a series of steps to determine the actual problem.

When diagnosing a fault it is useful for the technician to be aware of the circumstances that cause the control module to set a fault. There are a specific set of circumstances that cause the control module to set a fault. This information is found in the workshop manual. This information will allow the technician to further refine the diagnostics necessary to solve the problem.

Task E.3 Diagnose the causes of emissions or driveability problems resulting from failure of computerized engine controls with no diagnostic trouble codes (DTCs) stored.

A hesitation on acceleration can be experienced from a faulty throttle position sensor (TPS). A flat spot in the sensor range could be causing the signal voltage to momentarily drop. This is interpreted by the computer as a decrease in throttle position, when actually the vehicle is still accelerating. The computer may never set a diagnostic trouble code (DTC) based on this type of fault, because the voltage never varies above or below the diagnostic window.

Task E.4 Use a scan tool or DMM (Digital Multimeter) to inspect, test, adjust, and replace computerized engine control system sensors, powertrain control module (PCM), actuators, and circuits.

In order to control engine operation the control module must have a certain set of sensor inputs on which to make decisions. Once these inputs are received the control module processes the signals and decides which course of action to take. The control module then outputs signals to a series of actuators which in turn provide the engine with the things it needs to operate efficiently. In order to troubleshoot a system effectively, the technician must confirm that the sensor is transmitting the appropriate signal to the control module and the control module is transmitting the proper signal to the actuator.

With the wiring connectors disconnected from the engine coolant temperature (ECT) sensor and the powertrain control module (PCM), connect an ohmmeter from each sensor terminal to the PCM terminal to which the wire is connected. Both sensor wires should indicate less resistance than indicated by the manufacturer. If the wires have a higher resistance than specified, the wires or wiring connectors must be repaired. The ECT may be removed and placed in a container of water with an ohmmeter connected across the sensor terminals. A thermometer is then placed in the water. When the water is heated, the sensor resistance should decrease. If there is no decrease, replace the ECT.

Task E.5 Use and interpret digital multimeter (DMM) readings.

Some computer inputs, such as the O_2 sensor, produce a low voltage signal and a very low current flow. A digital voltmeter must be used to test the O_2 sensor voltage. An analog voltmeter draws more current than a digital voltmeter; O_2 sensor damage occurs when tested with an analog voltmeter.

Task E.6 Test, remove, inspect, clean, service, and repair or replace power and ground distribution circuits and connections.

The power distribution circuit is the power and ground circuits from the battery, through the ignition switch and fuses, to the individual circuits on the vehicle. Connections must be free of corrosion, as it adds unwanted resistance to current flow.

Task E.7 **Practice recommended precautions when handling static sensitive devices.**

Static sensitive components are shipped in an anti-static envelope. This envelope should not be opened before you are ready to install the component. Locate a good ground on the vehicle, and connect a grounding strap from you to the vehicle ground before installing the chip. Do not handle the chip unnecessarily, and do not move around on the vehicle seat when installing the chip.

Task E.8 **Diagnose driveability and emissions problems resulting from failures of interrelated systems (cruise control, security alarms, torque controls, suspension controls, traction controls, torque management, A/C, and similar systems).**

Today's vehicles have multiple computers with multiple functions. These computers have the ability to communicate with each other. Some sensor inputs are received by one computer, and the signal is forwarded to other computers. If this signal is not received, this can be interpreted as a lost signal and may cause output signal problems.

Task E.9 **Diagnose the causes of emissions or driveability problems resulting from computerized spark timing controls; determine needed repairs.**

The ignition module receives an input from a hall effect pickup or a variable reluctance sensor; this signal is used to fire the coil(s) on start up. The ignition module sends this signal to the powertrain control module (PCM), and the PCM interprets it as a revolutions per minute (rpm) input. This signal between the ignition module and PCM is a digital signal. The PCM then sends a varying digital signal back to the ignition module. The module uses this signal as a computed timing signal and fires the coil(s) based on this information.

Task E.10 **Verify, repair, and clear diagnostic trouble codes (DTCs).**

Prior to the introduction of OBD II, each manufacturer had its own method for erasing DTCs from the memory of a PCM. These procedures should always be followed. Normally, verification of the repair is done by operating the engine and the related system and checking to see if the operation triggered the DTC. If it did not, the problem was probably solved.

On OBD II equipped vehicles, the fail records and the freeze frame data for the DTC that was diagnosed should be reviewed and recorded. Then use a scan tool's clear DTCs or clear info functions to erase the DTCs from the PCM's memory. Operate the vehicle within the conditions noted in the fail records and/or the freeze frame data. Then monitor the status information for the specific DTC until the diagnostic test associated with that DTC runs.

F. Engines Electrical Systems Diagnosis and Repair (4 Questions)

1. Battery (1 Question)

Task F.1.1 **Test and diagnose emissions or driveability problems caused by battery condition and connections.**

Battery voltage is not needed to start the engine but it is also very important in stabilizing the voltage during engine operation. Once the engine is running, the vehicle's charging system supplies the voltage needed and restores the charge of the battery. As the charge voltage is provided to the system, it will fluctuate depending on the sensed need of the battery. Once the battery has the same voltage as the output of the charging system, the system voltage is leveled.

Many systems on today's vehicles are monitored by electronic components that send information to a control module or computer. This information is delivered as changes in voltage. Therefore precise voltage control is important to effective engine control management. When there is a large fluctuation in system voltage, sensor outputs may be wrong. These will tell the computer the wrong thing and driveability and other problems may result.

Poor system and component grounds are a common problem that occurs on today's vehicles.

Task F.1.2 Test and diagnose the cause(s) of abnormal key-off battery drain; determine needed repairs.

Most computers have a few milliamperes of current draw when they are not in operation. This current draw is called parasitic load. Since many vehicles today have several computers, this current draw may discharge the battery if the vehicle is not driven for several weeks. Connect a multimeter in series between the negative battery cable and ground; measure the amount of current draw in milliamps and compare it to manufacturer specifications.

2. Starting System (1 Question)

Task F.2.1 Perform starter current draw test; determine needed action.

Starter current draw testing is only performed on batteries with a specific gravity of 1.190 or greater. Several different testers can be used. If an analog tester is used, always check the mechanical zero on each meter and adjust as necessary. Be sure that all electrical loads are off and the doors are closed, as additional loads will cause additional draw. The ignition is disabled and the engine is cranked while observing the ammeter and voltmeter readings. High current draw and low cranking speed usually indicate a defective starter. High current draw may also be caused by internal engine problems. A low cranking speed and low current draw with high cranking voltage usually indicate excessive resistance in the starter circuit, such as in the cables and connections.

Task F.2.2 Perform starter circuit voltage drop tests; determine needed action.

The resistance in an electrical wire may be checked by measuring the voltage drop across the wire with normal current flow in the wire. To measure the voltage drop across the positive battery cable, connect the positive voltmeter lead to the positive cable at the battery, and connect the negative voltmeter lead to the other end of the positive battery cable at the starter solenoid. Disable the ignition system. Crank the engine. The voltage drop indicated on the meter should not exceed 0.5V. If the voltage reading is above this figure, the cable has excessive resistance. If the cable ends are clean and tight, replace the cable. Connect the positive voltmeter lead to the positive battery cable on the starter solenoid, and connect the negative voltmeter lead to the starting motor terminal on the other side of the solenoid. Leave the voltmeter on the lowest scale and crank the engine. If the voltage drop exceeds 0.3V, the solenoid disc and terminals have excessive resistance.

Task F.2.3 Inspect, test, and repair or replace components and wires in the starter control circuit.

When testing the starter control circuit, connect the positive voltmeter lead to the positive battery cable at the battery, and connect the negative voltmeter lead to the solenoid winding terminal on the solenoid. Leave the ignition system disabled and place the voltmeter selector on the lowest scale. If the voltage drop across the control circuit exceeds 1.5V while cranking the engine, individual voltage drop tests on control circuit components are necessary to locate the high resistance problem.

3. Charging System (2 Questions)

Task F.3.1 Test and diagnose charging system problems that cause an undercharge, overcharge, or a no-charge condition or engine performance problems; determine needed action.

If alternator output is zero, the alternator field circuit may be open. The most likely place for an open circuit in the alternator is at the sliprings and brushes. When the alternator output is normal and the amps output is zero, the fuse link between the alternator battery terminal and the positive battery cable may be open. If the alternator output is less than specified, always be sure that the belt and belt tension are satisfactory. When the belt tension is satisfactory and the alternator output is less than specified, the alternator is defective.

On some types of alternators there is a method of "full fielding" the unit. This technique will bypass the regulator circuit and full alternator output will be obtained. In this case if the alternator has full output, then the regulator or its circuits have failed.

Task F.3.2 Inspect, adjust, and replace alternator (generator) drive belts, pulleys, and fans.

A loose belt causes low alternator output and a discharged battery. A loose, dry, or worn belt may cause squealing and chirping noises during acceleration and cornering. Belt tension may be checked by measuring the belt deflection. Press on the belt with the engine stopped to measure the belt deflection. ½ in (12.7 mm) per foot (30.5 cm) of free span is usually acceptable.

Task F.3.3 Inspect, test, and repair or replace charging circuit connectors and wires.

Wires should be checked for burned or melted conditions. Connector ring terminals should be checked for loose retaining nuts, which cause high resistance or intermittent open circuits. An open circuit may be caused by a terminal that is backed out of the connector. Terminals that are bent or damaged may cause shorts or open circuits. An open circuit occurs when the terminal is crimped over the insulation instead of the wire core. A greenish white corrosion on terminals results in high resistance or an open circuit.

5 Sample Test for Practice

Sample Test

Please note the letter and number in parentheses following each question. They match the overview in section 4 that discusses the relevant subject matter. You may want to refer to the overview using this cross-referencing key to help with questions posing problems for you.

1. Technician A says the electrolyte level is not important in a non-serviceable battery. Technician B says the electrolyte level can be checked in a sealed battery by looking through the translucent battery case, on some batteries. Who is right?
   A. A only
   B. B only
   C. Both A and B
   D. Neither A nor B (F.1.1)
2. The LEAST likely cause of a fuel tank leak is:
   A. defective tank straps.
   B. road damage.
   C. defective seams.
   D. corrosion. (C.4)
3. Technician A says that nylon fuel piping can be formed around gradual curves. Technician B says that even in sharp bends, nylon pipe will allow fuel flow. Who is right?
   A. A only
   B. B only
   C. Both A and B
   D. Neither A nor B (C.4)
4. All of the following reduce turbocharger life EXCEPT:
   A. inadequate cooling.
   B. lack of oil changes.
   C. lack of air cleaner maintenance.
   D. exhaust system damage. (C.16)

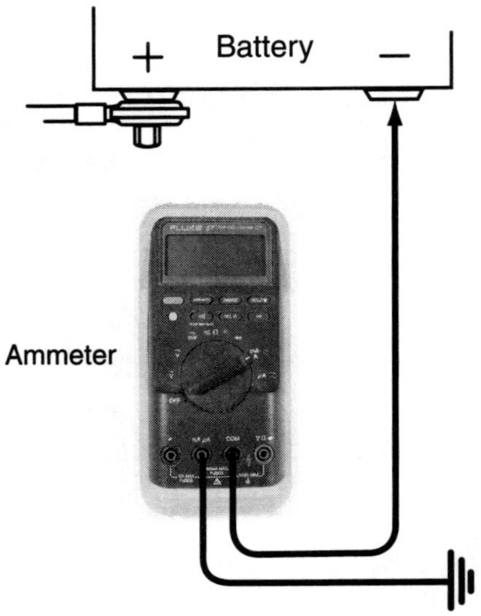

5. An ammeter, set in the milliamp position, is connected in series between the negative battery cable and ground, as shown. What is being measured?
   A. Starter draw
   B. Battery drain
   C. Regulated voltage
   D. Voltage drop (F.1.2)
6. Technician A says a fuel pressure test will test fuel pump operation. Technician B says it is possible to have a good pressure reading and insufficient flow. Who is right?
   A. A only
   B. B only
   C. Both A and B
   D. Neither A nor B (C.3)
7. Valve adjustment is being discussed. Technician A says valve adjustment should always be performed on a cold engine. Technician B says the piston should be placed at TDC of the compression stroke. Who is right?
   A. A only
   B. B only
   C. Both A and B
   D. Neither A nor B (All)

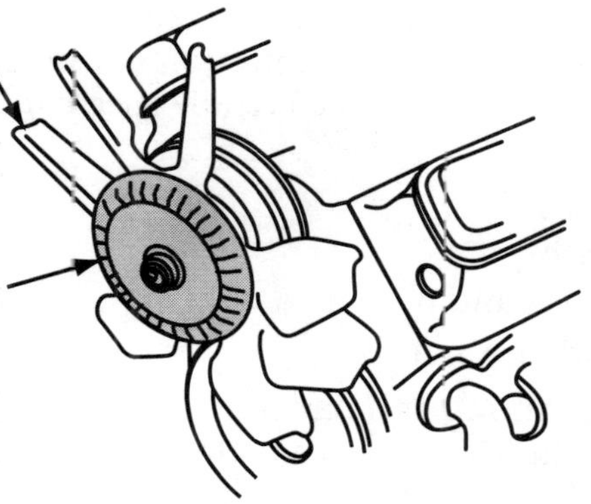

8. When testing a viscous-drive fan clutch, as shown in the figure, with the engine off, rotate the cooling fan by hand. It should have:
   A. more resistance hot.
   B. more resistance cold.
   C. no rotation movement.
   D. no resistance. (A.13)
9. All of the following are true of the cylinder leakage test EXCEPT:
   A. a gauge reading of 0 percent indicates no cylinder leakage.
   B. air loss from the PCV valve indicates worn piston rings.
   C. a gauge reading of 100 percent indicates no cylinder leakage.
   D. air loss from the exhaust indicates a valve problem. (A.8)
10. All of the following apply to exhaust manifold testing EXCEPT:
    A. checking for restricted exhaust manifold.
    B. cleaning all mating surfaces.
    C. checking for warping with a straightedge.
    D. inspecting for cracks. (C.14)
11. When installing and timing the distributor, Technician A says the engine must be timed referencing TDC on the specified cylinder's compression stroke. Technician B says if the engine is timed on the compression stroke, the distributor will be 180 degrees off. Who is right?
    A. A only
    B. B only
    C. Both A and B
    D. Neither A nor B (B.7)
12. While discussing positive crankcase ventilation (PCV) system diagnosis, Technician A says with the PCV valve disconnected from the rocker cover, there should not be vacuum at the valve with the engine idling. Technician B says when the PCV valve is removed and shaken, there should not be a rattling noise. Who is right?
    A. A only
    B. B only
    C. Both A and B
    D. Neither A nor B (D.1.2)
13. All of the following are true of ignition module replacement EXCEPT:
    A. dielectric silicone is used to seal the mating surface.
    B. replacement procedures vary depending upon application.
    C. dielectric silicone is used for heat dissipation.
    D. ignition module damage may occur without the use of dielectric silicone. (B.9)

14. Technician A says that idle mixture is set before idle speed and ignition timing. Technician B says an inspection of associated vacuum lines and electrical connectors should be done before idle speed and timing adjustments are attempted. Who is right?
    A. A only
    B. B only
    C. Both A and B
    D. Neither A nor B (C.12, C.13)

15. Technician A says a battery should be charged at one-half of the ampere-hour rating. Technician B says a battery should be charged until specific gravity is above 0.960. Who is right?
    A. A only
    B. B only
    C. Both A and B
    D. Neither A nor B (F.1.1)

16. Technician A says today's computer systems have the ability to communicate between multiple computers. Technician B says one input might affect several computers. Who is right?
    A. A only
    B. B only
    C. Both A and B
    D. Neither A nor B (E.8)

Ohmmeter

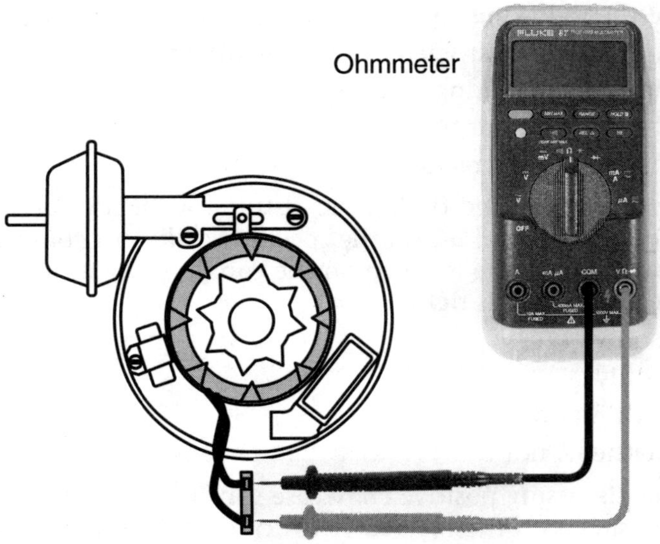

17. When testing a pickup coil with an ohmmeter, as shown in the figure, Technician A says when the pickup coil leads are moved, an erratic ohmmeter reading is normal. Technician B says that an infinite ohmmeter reading between the pickup coil terminals is an acceptable reading. Who is right?
    A. A only
    B. B only
    C. Both A and B
    D. Neither A nor B (B.8)

18. The ignition module uses the digital signal received from the PCM for:
    A. rpm input.
    B. hall effect timing.
    C. #1 cylinder signal.
    D. computed timing signal. (E.9)
19. While discussing engine performance diagnosis, Technician A says a vacuum leak decreases engine performance. Technician B says propane is the best method of locating vacuum leaks. Who is right?
    A. A only
    B. B only
    C. Both A and B
    D. Neither A nor B (C.11)
20. An engine has a lack of power and excessive fuel consumption. Technician A says a broken timing belt cannot be the cause. Technician B says the timing belt may have jumped a tooth. Who is right?
    A. A only
    B. B only
    C. Both A and B
    D. Neither A nor B (A.12)
21. Technician A says fuel reduction strategy is when a certain speed is reached and the computer shuts off the fuel injectors. Technician B says deceleration fuel reduction strategy is when the computer shuts off the injectors for a limited amount of time under deceleration. Who is right?
    A. A only
    B. B only
    C. Both A and B
    D. Neither A nor B (C.7)
22. When replacing a PROM, Technician A says that you should never ground yourself to the vehicle. Technician B says that grounding yourself to the vehicle will erase the PROM. Who is right?
    A. A only
    B. B only
    C. Both A and B
    D. Neither A nor B (E.7)
23. While discussing computers and input sensors, Technician A says a faulty throttle position sensor (TPS) can cause a hesitation on acceleration. Technician B says a faulty TPS will always set a diagnostic trouble code (DTC). Who is right?
    A. A only
    B. B only
    C. Both A and B
    D. Neither A nor B (E.2, E.3)
24. Technician A says some evaporative emissions canisters have a replaceable filter. Technician B says if the filler cap is equipped with pressure and vacuum valves, they must be checked for dirt contamination and damage. Who is right?
    A. A only
    B. B only
    C. Both A and B
    D. Neither A nor B (D.4.3)

25. When diagnosing fuel injection systems, a technical service bulletin search is performed for all of the following reasons EXCEPT:
    A. to save diagnostic time.
    B. year, make, and model specific information.
    C. to show all diagnostic specifications.
    D. mid-production changes. (A.2)
26. The LEAST likely symptom resulting from evaporative emission system failures is:
    A. rough idle.
    B. low emissions.
    C. malfunction indicator lamp (MIL) illumination.
    D. fuel odor. (D.4.1, D.4.2)
27. Technician A says the evaporative emission system assists in fuel vaporization in the intake manifold. Technician B says the evaporative emissions system prevents fuel vapors from escaping into the atmosphere. Who is right?
    A. A only
    B. B only
    C. Both A and B
    D. Neither A nor B (D.4.1)
28. Technician A says with a functional air injection reactor (AIR) pump disconnected and the engine running at 2,500 rpm under no load, the O_2 level should drop at least 2 percent. Technician B says the O_2 readings will not change with the AIR system disconnected. Who is right?
    A. A only
    B. B only
    C. Both A and B
    D. Neither A nor B (D.3.2)
29. If air is not exhausted from the secondary air injection bypass (AIRB) valve for a short time after startup, all of the following could be the problem EXCEPT:
    A. an AIRB solenoid.
    B. air supply from the pump to the valve.
    C. an AIRB valve.
    D. a one-way check valve. (D.3.3)
30. Technician A says if the air injection reactor (AIR) system pressure relief valve is stuck open, airflow is continually exhausted through the valve. Technician B says a relief valve that is stuck will cause high tailpipe emissions. Who is right?
    A. A only
    B. B only
    C. Both A and B
    D. Neither A nor B (D.3.3)

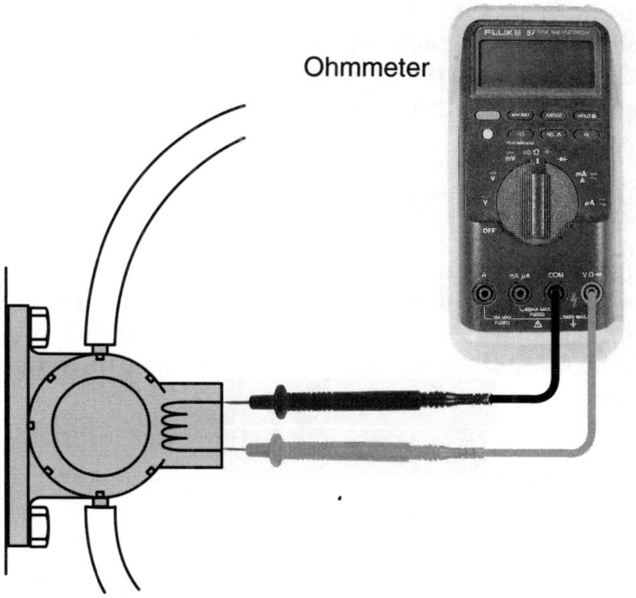

31. An EGR vacuum regulator solenoid (EGRV) is thought to be inoperative. Technician A says when an ohmmeter is connected, as shown, a lower than specified reading means the windings are open. Technician B says an infinite reading means the winding is shorted. Who is right?
    A. A only
    B. B only
    C. Both A and B
    D. Neither A nor B (D.2.3)
32. All of the following apply to an inoperative positive crankcase ventilation (PCV) system EXCEPT:
    A. plugged PCV valve.
    B. excessive blowby.
    C. kinked or restricted hoses.
    D. oil in the air cleaner housing. (D.1.2)
33. All of the following reduce turbocharger boost pressure EXCEPT:
    A. turbocharger bearing damage.
    B. wastegate stuck open.
    C. a leak between turbo and intake.
    D. inoperative engine coolant sensor. (C.16)
34. Technician A says restricted exhaust may cause reduced engine power. Technician B says restricted exhaust causes reduced maximum speed. Who is right?
    A. A only
    B. B only
    C. Both A and B
    D. Neither A nor B (C.15)

35. Technician A says if the positive crankcase ventilation (PCV) valve is stuck open, excessive airflow through the valve causes a rich air:fuel ratio. Technician B says if the PCV valve is restricted, excessive crankcase pressure forces blowby gases through the clean air hose into the air filter. Who is right?
    A. A only
    B. B only
    C. Both A and B
    D. Neither A nor B (D.1.1)

36. Technician A says the first step of any diagnostic procedure is to access computer system diagnostic trouble codes (DTCs). Technician B says that the first thing that should be done is a verification of the complaint. Who is right?
    A. A only
    B. B only
    C. Both A and B
    D. Neither A nor B (A.1)

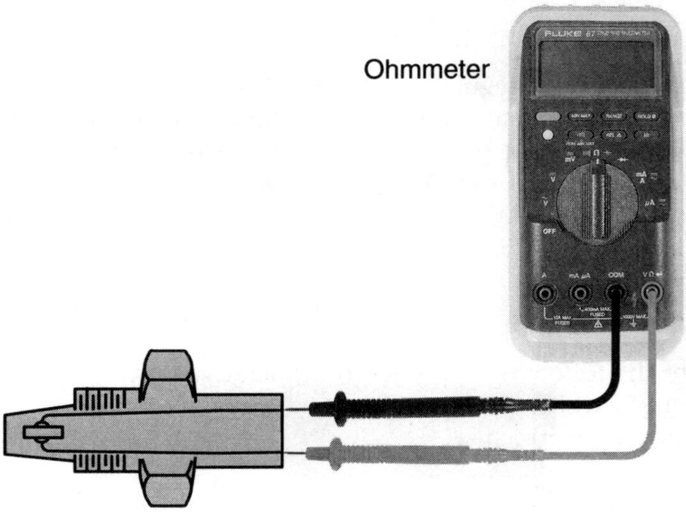

37. In the figure, the coolant sensor (a negative temperature coefficient type) is removed and placed in a container of water with an ohmmeter connected across the sensor terminals. Technician A says as the water temperature increases, the sensor resistance decreases. Technician B says as the water temperature increases, the sensor resistance increases. Who is right?
    A. A only
    B. B only
    C. Both A and B
    D. Neither A nor B (E.4, E.5)

38. Technician A says that when vacuum is applied to the exhaust gas recirculation (EGR) valve with the engine idling, the EGR valve should open and idle should become erratic. Technician B says that a diagnosis of the EGR valve should not be done with the engine idling. Who is right?
    A. A only
    B. B only
    C. Both A and B
    D. Neither A nor B (D.2.3)

39. While testing a pulse-air-type air injection system, Technician A says with the air cleaner clean air hose removed and the engine idling there should be steady audible pulses at the end of the hose. Technician B says if the pulses are erratic, a cylinder misfire might be the cause. Who is right?
    A. A only
    B. B only
    C. Both A and B
    D. Neither A nor B (D.3.1)

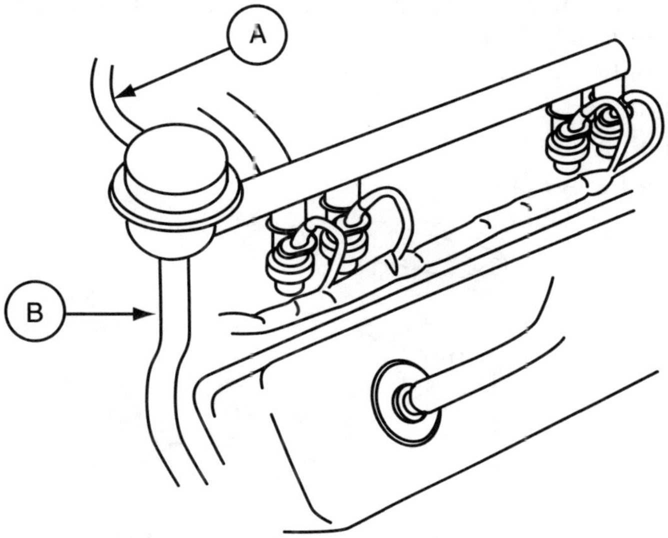

40. Technician A says by disconnecting the vacuum line to the pressure regulator during idle, as shown at A in the figure, a pressure increase of about 10 PSI (69 kPa) should be noted on the fuel gauge. Technician B says if fuel pressure increases after the return fuel line is restricted, at B, while doing a key-on, engine-off fuel pressure test, the pressure regulator is stuck open. Who is right?
    A. A only
    B. B only
    C. Both A and B
    D. Neither A nor B (C.6)
41. While testing the cooling system, Technician A says a repeated pressure test is the best way to ensure that all leaks are found. Technician B says a pressure test would also include the radiator cap. Who is right?
    A. A only
    B. B only
    C. Both A and B
    D. Neither A nor B (F.4)
42. If the vacuum drops slowly to a low reading when a vacuum gauge is connected to the intake manifold and the engine is accelerated and held at a steady speed, it indicates:
    A. sticking valves.
    B. over-advanced ignition timing.
    C. a restricted exhaust.
    D. a rich fuel mixture. (A.5, C.16)

43. Technician A says a defective MAP sensor may cause a rich or lean air:fuel ratio. Technician B says by measuring the MAP signal under different operating conditions, the MAP sensor can be accurately diagnosed. Who is right?
   A. A only
   B. B only
   C. Both A and B
   D. Neither A nor B (C.1, E.4)

44. Technician A says a greenish white corrosion on terminals results in high resistance. Technician B says loose retaining nuts cause high resistance in connector ring terminals. Who is right?
   A. A only
   B. B only
   C. Both A and B
   D. Neither A nor B (E.5)

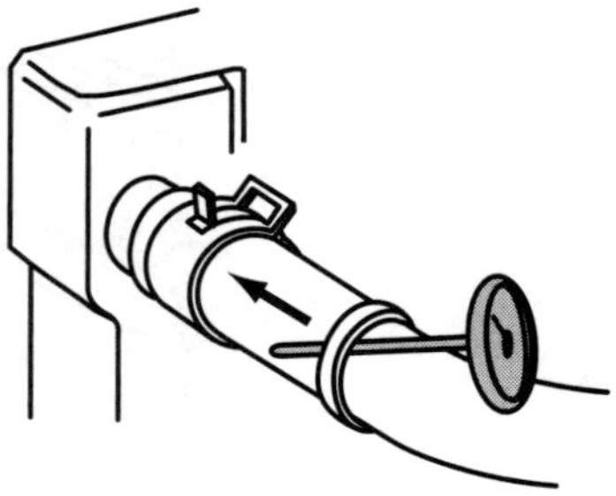

45. With a thermometer taped to the upper radiator hose, as shown in the figure, and the vehicle idling for fifteen minutes, the temperature indication should be:
   A. lower than the thermostat-specified temperature.
   B. higher than the thermostat-specified temperature.
   C. within a few degrees of the thermostat-specified temperature.
   D. one-half of the thermostat-specified temperature. (A.13)

46. When testing a catalytic converter for restriction, Technician A says a loss of power and a limited top speed will be noticed. Technician B says if you tap on certain types of converters with a hammer and the converter rattles, it should be replaced. Who is right?
   A. A only
   B. B only
   C. Both A and B
   D. Neither A nor B (D.3.4)

47. Technician A says a diagnostic trouble code (DTC) tells you what component should be replaced. Technician B says if a fault exists that affects emissions, the emission reminder lamp will be illuminated. Who is right?
   A. A only
   B. B only
   C. Both A and B
   D. Neither A nor B (E.2)

48. The LEAST likely problem resulting from an accelerator pump out of adjustment is:
   A. hesitation.
   B. stumble on acceleration.
   C. lean misfire at cruise speeds.
   D. excess fuel consumption. (C.7)

49. Technician A says an analog voltmeter should be used for O_2 sensor voltage checks. Technician B says analog voltmeters are typically of a low impedence input design, and cannot be used, because they draw too much current and can damage the O_2 sensor. Who is right?
   A. A only
   B. B only
   C. Both A and B
   D. Neither A nor B (E.5)

50. A test lamp is connected between the negative side of the coil and ground to diagnose a no-start condition. Technician A says a flickering test lamp indicates a defective ignition module. Technician B says a flickering test lamp indicates a defective pickup coil. Who is right?
   A. A only
   B. B only
   C. Both A and B
   D. Neither A nor B (B.1)

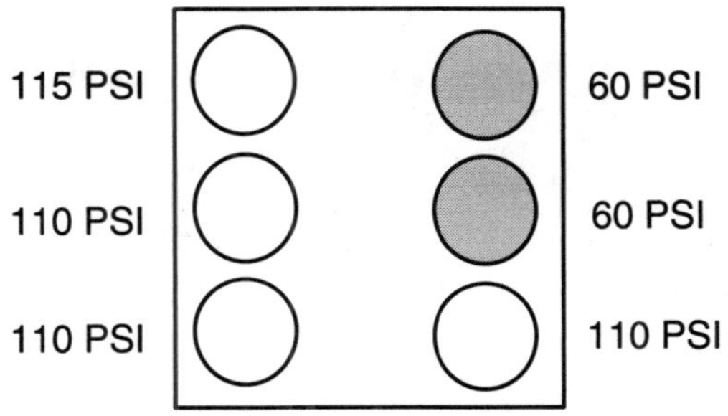

51. Technician A says that during a cylinder compression test, low readings on adjacent cylinders, as shown in the figure, may be caused by a blown cylinder head gasket. Technician B says a low reading on one cylinder is probably a piston ring or valve problem. Who is right?
   A. A only
   B. B only
   C. Both A and B
   D. Neither A nor B (A.7)

52. The first thing a technician should do to test a cold start injector is to:
   A. check the spray pattern.
   B. perform injector balance testing.
   C. check the cold start injector resistance value.
   D. energize the injector and watch for a pressure drop. (C.7)

53. When voltage drop testing the power distribution and ground circuits, Technician A says all connections must be free and clean of corrosion. Technician B says corrosion adds unwanted resistance. Who is right?
   A. A only
   B. B only
   C. Both A and B
   D. Neither A nor B (E.6)

54. A cylinder power balance test can indicate all of the following problems EXCEPT:
   A. a bad spark plug.
   B. late ignition timing.
   C. an open ignition wire.
   D. burned valves. (A.6)

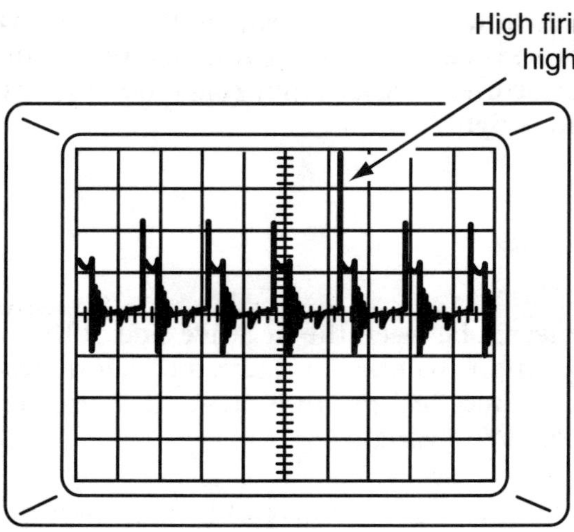

55. While monitoring secondary ignition with an oscilloscope, as shown in the figure, the LEAST likely cause of high resistance in the ignition secondary circuit is:
    A. damaged carbon spark plug wires.
    B. no dielectric compound on the ignition module mounting surface.
    C. corroded spark plug wire ends.
    D. excessive rotor air gap. (A.9, B.1, B.5)
56. The LEAST likely test performed with an emission analyzer is:
    A. cylinder head gasket leak.
    B. O_2 sensor waveforming.
    C. cylinder misfiring.
    D. inspection and maintenance program exhaust analysis. (A.10)

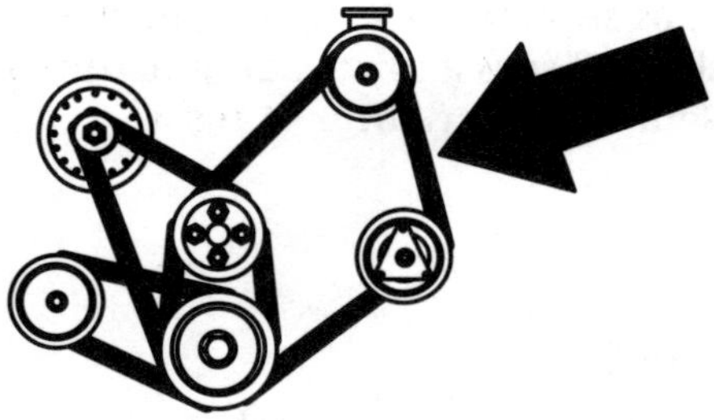

57. Referring to the figure, Technician A says a deteriorated belt could cause squealing and chirping. Technician B says belt deflection should be under ½ in (12.7 mm) per foot (30.5 cm) free span. Who is right?
    A. A only
    B. B only
    C. Both A and B
    D. Neither A nor B (F.3.2)
58. While performing starter current draw testing, all of the following are true EXCEPT:
    A. the ignition must be disabled.
    B. the battery must be fully charged.
    C. the doors are closed.
    D. electrical loads are on. (F.2.1)

59. Technician A says that an oscilloscope can be used to watch the O² sensor signal switch from rich to lean status. Technician B says fuel injectors can only be tested using the fuel injector balance tester. Who is right?
    A. A only
    B. B only
    C. Both A and B
    D. Neither A nor B (A.9, E.4)
60. The LEAST likely problem with an air injection reactor (AIR) pump is:
    A. misrouted air.
    B. a restricted filter.
    C. a bent pulley.
    D. a worn pump shaft. (D.3.3)
61. Technician A says timing specifications can be found on the underhood emission label. Technician B says the timing light can be connected to any plug wire to obtain the proper firing input. Who is right?
    A. A only
    B. B only
    C. Both A and B
    D. Neither A nor B (B.7)
62. Which of the following is NOT an approved way to obtain diagnostic trouble codes (DTC)?
    A. Connecting jumper wires to the data link connector.
    B. Cycling the malfunction indicator lamp (MIL) on and off.
    C. Cycling the ignition switch three times in five seconds.
    D. Using a scan tool. (E.1)
63. Technician A says if the exhaust gas recirculation (EGR) valve remains open at idle and low speed, the idle will be rough. Technician B says if the EGR valve does not open, detonation can occur. Who is right?
    A. A only
    B. B only
    C. Both A and B
    D. Neither A nor B (D.2.1)
64. Technician A says a voltage drop test checks the amount of resistance between two test points. Technician B says more than a 0.5V voltage drop indicates excessive resistance across the battery positive cable. Who is right?
    A. A only
    B. B only
    C. Both A and B
    D. Neither A nor B (F.1.1)
65. Technician A says in dusty conditions, a damaged air filter can increase wear on cylinder walls. Technician B says an air filter problem does not affect fuel consumption. Who is right?
    A. A only
    B. B only
    C. Both A and B
    D. Neither A nor B (C.10, C.11)

66. While disassembling and testing a turbocharger compressor wheel or turbine wheel housing, scoring is observed. Technician A says turbine wheel housing scoring is caused by excessive turbocharger shaft end play. Technician B says turbine wheel housing scoring is caused by a leak in the air intake. Who is right?
    A. A only
    B. B only
    C. Both A and B
    D. Neither A nor B (C.16)
67. Technician A says replacement with a known good part is the only way to test an ignition module. Technician B says an ignition module tester can give a good/bad indication. Who is right?
    A. A only
    B. B only
    C. Both A and B
    D. Neither A nor B (B.9)

68. While discussing throttle body carbon deposits, as shown in the figure, Technician A says if an approved aerosol spray cleaner doesn't properly clean a throttle body, the throttle body must be removed and disassembled. Technician B says a buildup of gum and carbon deposits may cause rough idle operation. Who is right?
    A. A only
    B. B only
    C. Both A and B
    D. Neither A nor B (C.9)
69. Technician A says because of extreme temperatures in the exhaust stream, restricted EGR passages are not a problem. Technician B says that before any EGR parts are replaced, all passages should be clean. Who is right?
    A. A only
    B. B only
    C. Both A and B
    D. Neither A nor B (D.2.3)
70. When the belt and belt tension are satisfactory and the alternator output is low, Technician A says the alternator may be defective. Technician B says the problem could be the slip rings and the brushes. Who is right?
    A. A only
    B. B only
    C. Both A and B
    D. Neither A nor B (F.3.1)

71. Technician A says when testing the starter control circuit, the voltage drop during cranking should not exceed 3.5 volts. Technician B says individual voltage drop tests on control circuit components may be necessary. Who is right?
   A. A only
   B. B only
   C. Both A and B
   D. Neither A nor B (F.2.2)

72. All of the following are necessary to complete the injector cleaning process EXCEPT:
   A. a solution of unleaded fuel and injector cleaner.
   B. resetting the adaptive memory.
   C. activating the vehicle's fuel pump.
   D. restricting the fuel return line. (C.9)

73. While testing fuel pressure on a TBI engine, Technician A says there will always be a Shrader test port for fuel system testing. Technician B says that a high fuel pressure reading could be the result of a plugged fuel filter trapping fuel between the filter and the fuel rail. Who is right?
   A. A only
   B. B only
   C. Both A and B
   D. Neither A nor B (C.3)

74. Technician A says a battery capacity tester that houses a carbon pile is a variable load tester. Technician B says a carbon pile is a fixed load tester. Who is right?
   A. A only
   B. B only
   C. Both A and B
   D. Neither A nor B (F.1.1)

75. While discussing exhaust color diagnosis, Technician A says black smoke in the exhaust indicates a lean air-fuel mixture. Technician B says gray smoke in the exhaust indicates oil leakage into the combustion chamber. Who is right?
   A. A only
   B. B only
   C. Both A and B
   D. Neither A nor B (A.4)

76. While performing engine noise diagnostics, Technician A says a stethoscope is a good tool for noise location. Technician B says you must duplicate the specific operating condition. Who is right?
   A. A only
   B. B only
   C. Both A and B
   D. Neither A nor B (A.3)

77. Technician A says a thorough ignition coil test includes both primary and secondary winding resistance tests. Technician B says maximum coil output testing can be performed with an oscilloscope. Who is right?
   A. A only
   B. B only
   C. Both A and B
   D. Neither A nor B (B.6)

6 Additional Test Questions for Practice

Additional Test Questions

Please note the letter and number in parentheses following each question. They match the overview in section 4 that discusses the relevant subject matter. You may want to refer to the overview using this cross-referencing key to help with questions posing problems for you.

1. A vehicle emits a belt squeal while cornering and on acceleration. Technician A says the alternator bearings may be defective. Technician B says that the alternator belt may be loose. Who is right?
   A. A only
   B. B only
   C. Both A and B
   D. Neither A nor B (F.3.2)
2. A diverter valve in the AIR system is used to:
   A. prevent backfire on deceleration.
   B. enrich fuel mixture on deceleration.
   C. divert cold air into the passenger compartment.
   D. turn the A/C compressor off. (D.3.1)
3. Technician A says that diagnostic trouble codes (DTCs) can be retrieved by using a scan tool. Technician B says in some cases DTCs can be retrieved using an analog voltmeter. Who is right?
   A. A only
   B. B only
   C. Both A and B
   D. Neither A nor B (E.1)

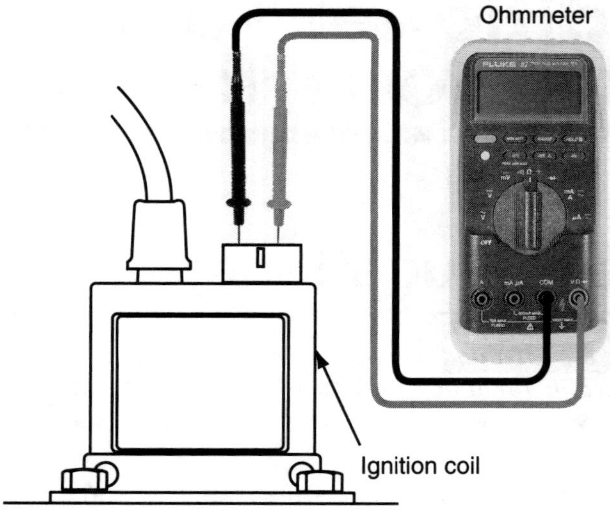

4. Technician A says the ohmmeter in the figure reads infinity and means the circuit has little or no resistance. Technician B says this means a circuit is open. Who is right?
   A. A only
   B. B only
   C. Both A and B
   D. Neither A nor B (E.5)
5. Technician A says you can use a stethoscope to locate a plugged PCV valve. Technician B says you need a vacuum gauge to properly diagnose a PCV system. Who is right?
   A. A only
   B. B only
   C. Both A and B
   D. Neither A nor B (D.1.1)

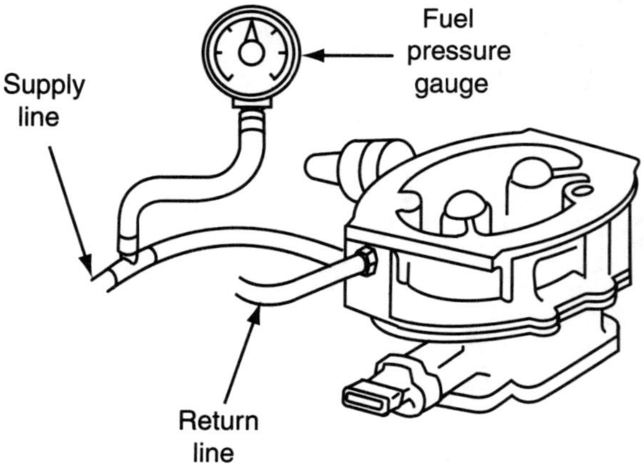

6. In the figure, Technician A says the throttle body must be completely disassembled before soaking it in cleaning solution. Technician B says the test being performed is for fuel pressure. Who is right?
   A. A only
   B. B only
   C. Both A and B
   D. Neither A nor B (C.8)

7. While applying dielectric silicone to an ignition module mounting surface, Technician A says dielectric silicone is used to dissipate heat. Technician B says without dielectric silicone, the ignition module may experience overheat conditions. Who is right?
   A. A only
   B. B only
   C. Both A and B
   D. Neither A nor B (B.9)

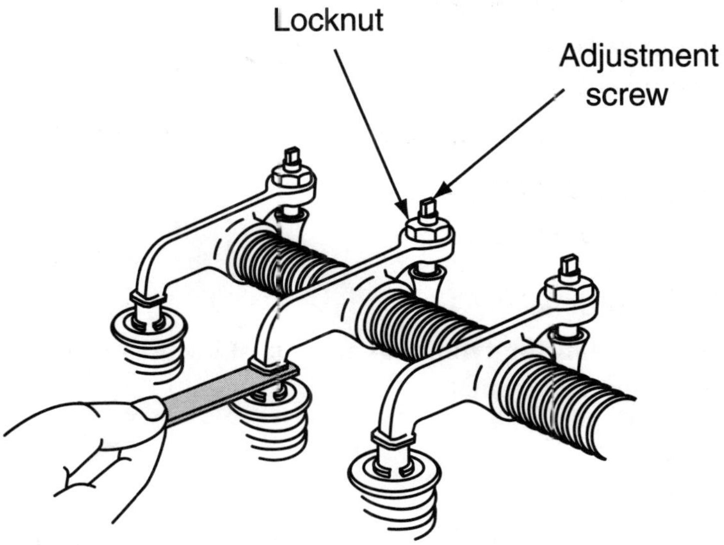

8. If the measurement in the figure is set too wide, Technician A says it will retard valve timing. Technician B says it will reduce valve overlap. Who is right?
   A. A only
   B. B only
   C. Both A and B
   D. Neither A nor B (A.11)
9. The LEAST likely cause of low cylinder compression is:
   A. worn valves.
   B. worn rings.
   C. blown head gasket.
   D. worn valve guides. (A.7)
10. The result of a battery load test done with a carbon pile load tester is 8.7 volts. Technician A says this is unacceptable. Technician B says that you should compare the results to the tables from the tool manufacturer. Who is right?
    A. A only
    B. B only
    C. Both A and B
    D. Neither A nor B (F.1.1)
11. A double knocking noise is heard from an engine while idling. Technician A says this could be from worn piston pins. Technician B says this could also be caused by worn main bearings. Who is right?
    A. A only
    B. B only
    C. Both A and B
    D. Neither A nor B (A.3)

12. When a vacuum leak is suspected for a rough idle complaint, the best check would be to use:
    A. carburetor cleaner.
    B. propane.
    C. water.
    D. a stethoscope. (C.11)
13. The LEAST likely cause of blue exhaust emissions is:
    A. worn piston rings.
    B. damaged cylinder head.
    C. worn valve seals.
    D. worn valve seats. (A.4)
14. A carbureted vehicle has a tip-in hesitation when warm. Which of these should be checked first?
    A. Stuck open choke
    B. Accelerator pump
    C. Fuel filter
    D. Fuel return line (C.7)
15. All of the following symptoms are true of low fuel pressure EXCEPT:
    A. excessive sulfur smell.
    B. lack of power.
    C. engine surging.
    D. limited top speed. (C.1)
16. A 12-volt battery has just failed a capacity test and is being charged at 40 amps. After three minutes of charge, with the charger still operating, a voltmeter is hooked across the battery and reads 15.8 volts. What does this indicate?
    A. The battery should be slow charged and put back into service.
    B. The battery's electrolyte should be replaced.
    C. The battery is sulfated and should be replaced.
    D. This is normal; continue fast charge and return to service. (F.1.1)

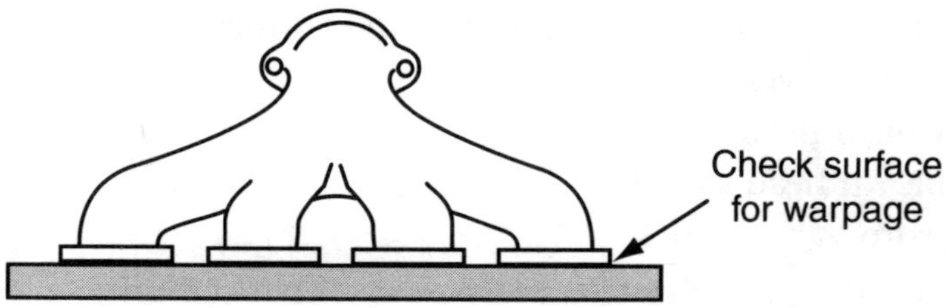

17. To check an exhaust manifold for warpage between ports, as shown in the figure, Technician A says that only a straightedge is needed. Technician B says a straightedge and a flashlight should be used. Who is right?
    A. A only
    B. B only
    C. Both A and B
    D. Neither A nor B (C.14)
18. The first symptom of a restricted air filter is:
    A. poor fuel economy.
    B. a no-start condition.
    C. excessive engine wear.
    D. excessive oil consumption. (C.10)

19. The best test for positive crankcase ventilation (PCV) operation is to:
    A. shake the PCV valve and listen for a rattle.
    B. place a finger over inlet side of the PCV valve and feel for vacuum.
    C. take a reading with the vacuum gauge on the outlet side.
    D. inspect for oil residue on either side of the valve. (D.1.2)

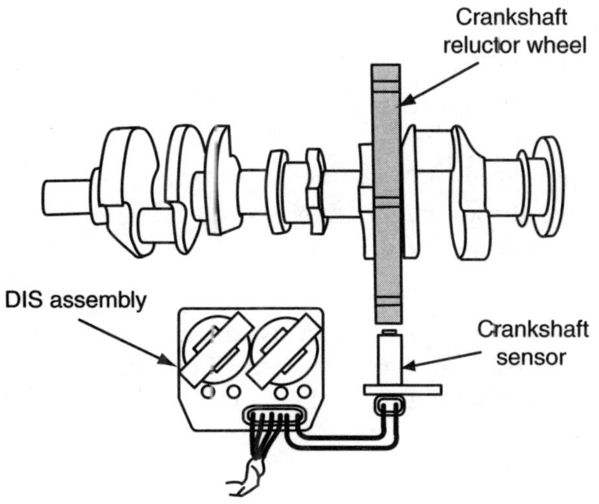

20. An engine equipped with a distributorless ignition system as shown won't start. Technician A says this could be caused by a defective crankshaft sensor. Technician B says this could be caused by a faulty signal from the camshaft position sensor. Who is right?
    A. A only
    B. B only
    C. Both A and B
    D. Neither A nor B (B.1)
21. Technician A says that the ignition module controls the timing on start-up only. Technician B says that because some systems have no distributor, the powertrain control module (PCM) has full control of timing at all times. Who is right?
    A. A only
    B. B only
    C. Both A and B
    D. Neither A nor B (B.7)
22. A technician is performing a compression test. Which statement below is LEAST likely true?
    A. All cylinders with higher than normal readings could be caused by carbon buildup.
    B. All cylinders reading even, but lower than normal, may be caused by a slipped timing chain.
    C. Low readings on two adjacent cylinders may be caused by a blown head gasket.
    D. A low reading on one cylinder may be caused by a vacuum leak at that cylinder. (A.7)
23. A turbocharger requires frequent replacement due to bearing failure. Technician A says restricted coolant passages to the turbocharger could be the problem. Technician B says restricted oil passages to the turbocharger could be the problem. Who is right?
    A. A only
    B. B only
    C. Both A and B
    D. Neither A nor B (C.16)

24. While testing a turbocharger, the maximum boost pressure observed is 4 PSI (27.6 kPa), while the specified pressure is 9 PSI (62 kPa). Technician A says the engine compression may be low. Technician B says the wastegate may be sticking open. Who is right?
    A. A only
    B. B only
    C. Both A and B
    D. Neither A nor B (C.16)
25. The vanes on a turbocharger compressor wheel are severely pitted. The cause of this problem could be:
    A. partially seized turbocharger bearings.
    B. excessive turbocharger shaft end play.
    C. a leak in the air intake system.
    D. an overheated turbocharger. (C.16)
26. Technician A says that you can use a stethoscope to locate a plugged positive crankcase ventilation (PCV) system. Technician B says that you need a vacuum gauge to properly diagnose a PCV system. Who is right?
    A. A only
    B. B only
    C. Both A and B
    D. Neither A nor B (D.1.1)

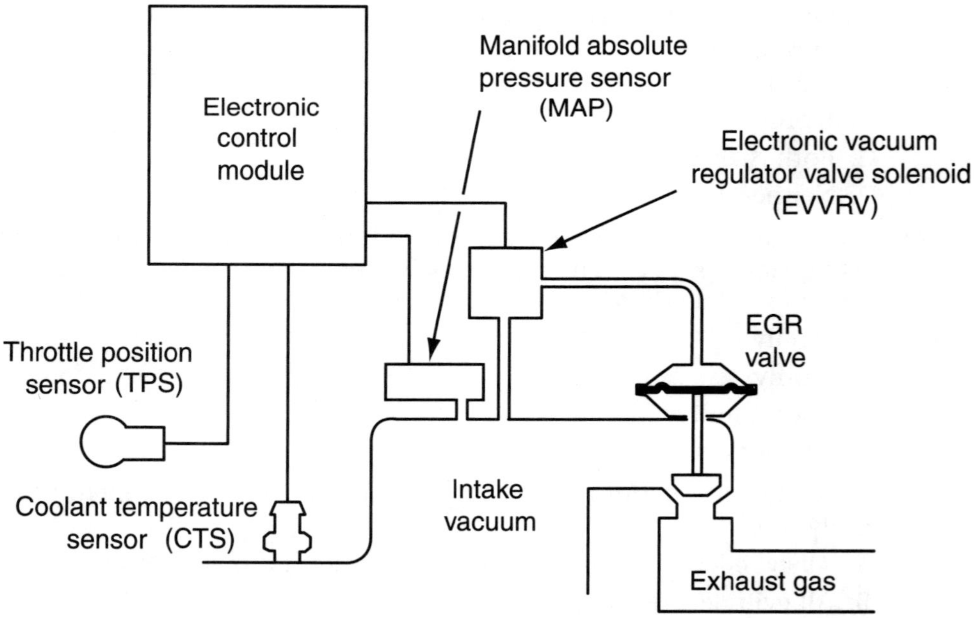

27. Technician A says that the first step in diagnosing any exhaust gas recirculation (EGR) valve concern is to check the vacuum and electrical connections. Technician B says that in many systems, as shown in the figure, the powertrain control module (PCM) uses other sensor inputs that could cause an EGR problem, and therefore diagnostic trouble codes (DTCs) should be corrected before replacing any EGR components. Who is right?
    A. A only
    B. B only
    C. Both A and B
    D. Neither A nor B (D.2.2)

28. A multiport fuel injected, distributor-type ingnition V-6 engine is running rough and has a lean fuel mixture. Injecting propane improves engine idle quality. There are no vacuum leaks. After performing a cylinder balance test, cylinders #1 and #5 are found to be weak. Technician A says this could be caused by a bad ignition module. Technician B says this could be caused by the #3 injector having low resistance. Who is right?
    A. A only
    B. B only
    C. Both A and B
    D. Neither A nor B (A.2, A.6, C.1, C.9)

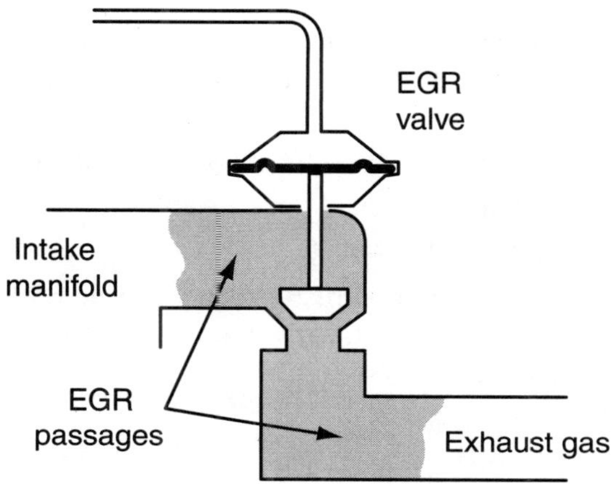

29. Technician A says to always check the exhaust passages when replacing an exhaust gas recirculation (EGR) valve, as shown in the figure. Technician B says an EGR valve that tests bad is inoperative; no other checks are needed. Who is right?
    A. A only
    B. B only
    C. Both A and B
    D. Neither A nor B (D.2.2, D.2.3)

30. Technician A says that a DMM/DVOM can be used to check an oxygen sensor. Technician B says that to check an oxygen sensor you need a diode tester. Who is right?
    A. A only
    B. B only
    C. Both A and B
    D. Neither A nor B (E.4)

31. A proper cooling system inspection involves all of the following EXCEPT a:
    A. pressure test.
    B. recovery tank inspection.
    C. thermostat inspection.
    D. heater core inspection. (A.13)

32. High hydrocarbon (HC) emissions may be caused by all the following EXCEPT:
    A. a cylinder misfire.
    B. an excessive lean condition.
    C. a faulty fuel pressure regulator.
    D. power relay. (A.10)

33. Technician A says nylon fuel piping may be used to go around gradual curves. Technician B says nylon fuel piping will allow fuel flow in sharp bends. Who is right?
   A. A only
   B. B only
   C. Both A and B
   D. Neither A nor B (C.4)

34. The LEAST likely cause of poor fuel mileage on a vehicle with EFI is:
   A. high fuel pressure.
   B. disconnected regulator vacuum hose.
   C. partially plugged exhaust.
   D. low fuel pressure. (C.3)

35. When the engine is started, air should be exhausted from the secondary air injection bypass (AIRB) valve for a short period of time. If this is not observed:
   A. check the fuse.
   B. remove the vacuum hose and retest, include checking for DTC.
   C. inspect the check valves.
   D. check for restricted exhaust. (D.3.2, D.3.3)

36. Technician A says carbon canister filters can be replaced. Technician B says filler caps with pressure and vacuum valves must be checked for contamination and damage. Who is right?
   A. A only
   B. B only
   C. Both A and B
   D. Neither A nor B (C.4, D.4.3)

37. A starter free-running test is being made on the bench with a fully charged battery. The current draw is higher than specification and the rpm is lower. Technician A says this could be caused by tight bushings. Technician B says this could be caused by worn brushes. Who is right?
   A. A only
   B. B only
   C. Both A and B
   D. Neither A nor B (F.2.1)

38. A slipped timing belt can cause all EXCEPT:
   A. poor fuel mileage.
   B. no-start.
   C. high manifold vacuum
   D. low power. (A.12)

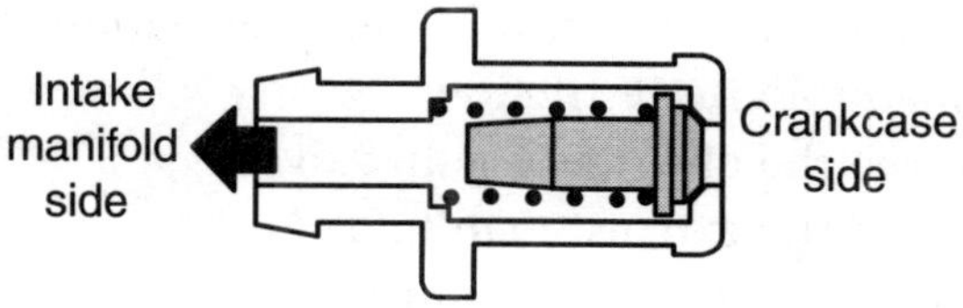

39. In the figure, the plunger is stuck in the maximum flow position. Technician A says this can cause a rough idle. Technician B says this could cause excessive oil consumption. Who is right?
   A. A only
   B. B only
   C. Both A and B
   D. Neither A nor B (D.1.1, D.1.2)

40. An evaporative emissions charcoal canister failure can be caused by all EXCEPT:
   A. injector failure.
   B. powertrain control module (PCM).
   C. vacuum leak.
   D. solenoid failure. (D.4.1, D.4.2)

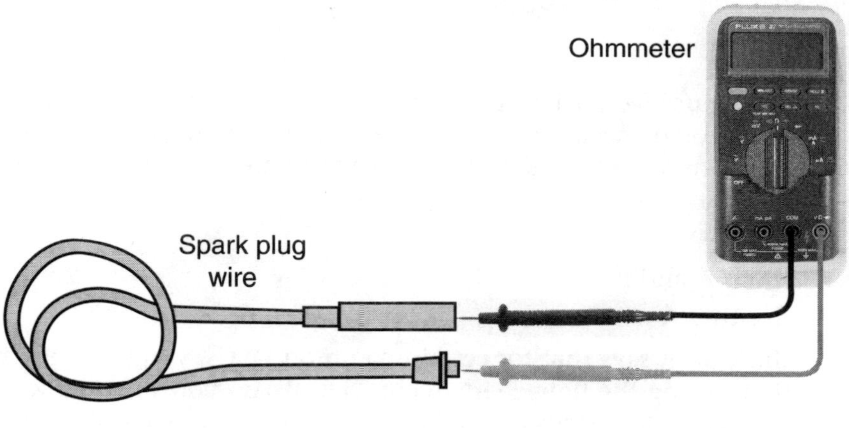

41. While testing a spark plug wire, as shown in the figure, Technician A says the manufacturer has a specified amount of resistance for each foot of plug wire. Technician B says plug wire resistance readings should indicate no (zero) resistance. Who is right?
   A. A only
   B. B only
   C. Both A and B
   D Neither A nor B (B.5)

42. All of the following statements about PCM inputs are true EXCEPT:
   A. digital voltmeters/ohmmeters may be used for diagnosis.
   B. some inputs, but not all, are very low voltage.
   C. with practice, an experienced technician can use an analog meter.
   D. the O_2 sensor produces very low voltage. (E.5)

43. A car with an AIR system backfires on deceleration. The technician should check:
    A. the air manifold for restrictions.
    B. operation of the diverter or gulp valve.
    C. operation of the exhaust manifold check valve.
    D. output pressure of the air pump. (D.3.1)

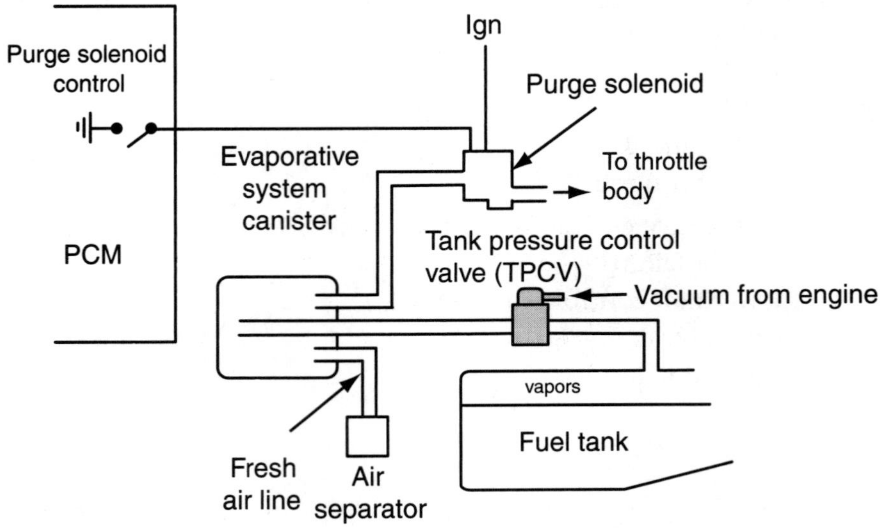

44. In the figure, Technician A says with a fuel tank pressure control valve off the vehicle, you could check the valve with a vacuum pump. Technician B says the valve must be installed on the vehicle, and a vacuum gauge must be used. Who is right?
    A. A only
    B. B only
    C. Both A and B
    D. Neither A nor B (D.4.3)
45. Technician A says that for certain common DTCs, you can replace the component without using the flow chart. Technician B says you should always use the flow chart, but certain steps in the flow chart can be skipped. Who is right?
    A. A only
    B. B only
    C. Both A and B
    D. Neither A nor B (E.2, E.10)
46. Technician A says if you are testing an ignition related no-start problem, you should always check for available spark at an ignition wire first. Technician B says if you have a test light connected to the negative side of the coil while cranking and the test light flickers, you need to test the secondary ignition system. Who is right?
    A. A only
    B. B only
    C. Both A and B
    D. Neither A nor B (B.1)

47. Technician A says the vacuum lines at the distributor are usually disconnected and plugged to check base ignition timing. Technician B says the distributor rotor must point at the specified cylinder's distributor cap terminal when installing the distributor. Who is right?
    A. A only
    B. B only
    C. Both A and B
    D. Neither A nor B (B.4, B.7)

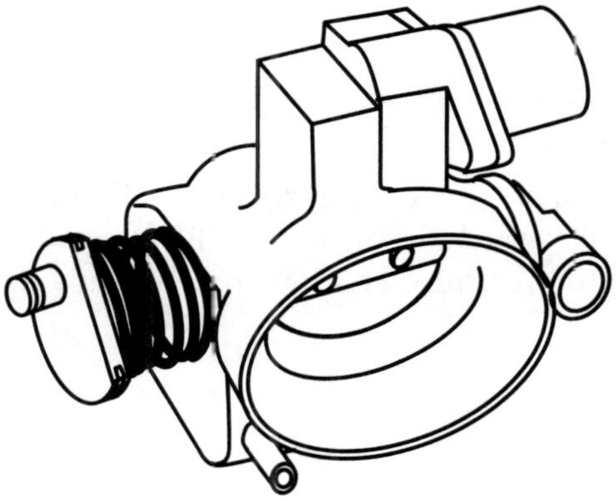

48. Technician A says the throttle body shown in the figure must be removed to be cleaned. Technician B says the minimum airflow rate and throttle plate angle needs to be checked and adjusted if necessary after cleaning. Who is right?
    A. A only
    B. B only
    C. Both A and B
    D. Neither A nor B (C.9, E.10)
49. A spark knock from the engine on medium to hard acceleration would most likely be caused by:
    A. a rich fuel mixture.
    B. a bad crankshaft sensor.
    C. high manifold vacuum.
    D. a restricted EGR. (D.2.1)
50. A technical service bulletin should be used for all of the following EXCEPT:
    A. to save diagnostic time.
    B. to show mid-production changes.
    C. to show sales information.
    D. to show year, make, and model information. (A.2)
51. During hot weather a vehicle fails an I/M 240 test with high CO at an idle. Which of the following would most likely be the cause?
    A. Constant high O_2 sensor voltage
    B. An EGR valve not seating
    C. A bad ignition module
    D. An EVAP purge solenoid stuck open (A.10, B.4.1)

52. Nylon fuel hoses should be inspected for all EXCEPT:
    A. kinks.
    B. loose fittings.
    C. discoloration.
    D. scratches. (C.14)
53. During a cylinder power balance test, there is no rpm drop on cylinder #3. Technician A says that the cylinder is not contributing. Technician B says that the cylinder may have an inoperative spark plug. Who is right?
    A. A only
    B. B only
    C. Both A and B
    D. Neither A nor B (A.5, B.1)
54. Technician A says ignition timing is usually set with the engine running and the computer power feed disconnected. Technician B says ignition timing is usually set with the engine running at 2500 rpm. Who is right?
    A. A only
    B. B only
    C. Both A and B
    D. Neither A nor B (B.7)
55. Technician A says you can determine that the electrolyte is low on any battery. Technician B says you can only check the electrolyte level accurately if it has fill caps. Who is right?
    A. A only
    B. B only
    C. Both A and B
    D. Neither A nor B (F.1.1)
56. All of the following are true about the complaint verification process EXCEPT:
    A. isolate the system causing the complaint.
    B. identify conditions under which the concern occurs.
    C. road test.
    D. review the repair order. (A.1)
57. Technician A says you will not harm the PCM with static electricity if the negative battery cable is disconnected. Technician B says you should never ground yourself to the vehicle while working on a PCM. Who is right?
    A. A only
    B. B only
    C. Both A and B
    D. Neither A nor B (E.7)
58. When testing voltage drop while cranking the engine with the ignition disabled, Technician A says the voltage drop on the positive battery cable should be no more than 5 volts. Technician B says the voltage drop on the positive battery cable should be no more than 1 volt. Who is right?
    A. A only
    B. B only
    C. Both A and B
    D. Neither A nor B (F.2.2)

59. The LEAST likely cause of ignition coil failure is:
    A. prolonged open circuit in the secondary.
    B. overheating the coil.
    C. coil case cracking.
    D. open ignition module primary circuit. (B.6)
60. A multiport fuel injected vehicle has poor fuel economy, yet starts and runs fine. Technician A says the fuel return line may be restricted. Technician B says the fuel pressure regulator may be stuck. Who is right?
    A. A only
    B. B only
    C. Both A and B
    D. Neither A nor B (C.6)
61. A restricted exhaust will cause vacuum readings to:
    A. drop off about three inches at an idle.
    B. drop off about eight inches at an idle.
    C. fluctuate between sixteen and twenty-one inches at an idle.
    D. show a continuous gradual drop as engine speed is increased. (A.5, C.15)
62. Many cooling systems use a thermal control fan drive coupling (fan clutch). Technician A says the thermostatic coil controls the opening and closing of the orifice inside the coupling. Technician B says when the thermostatic coil is cold, the orifice is open. Who is right?
    A. A only
    B. B only
    C. Both A and B
    D. Neither A nor B (A.14)
63. Technician A says performing a fuel pressure test will test the fuel pump operation. Technician B says it is possible to have a hydraulic problem with an injector, even though the electrical resistance is within specifications. Who is right?
    A. A only
    B. B only
    C. Both A and B
    D. Neither A nor B (C.3, C.9)
64. Technician A says an ignition module tester proves the module's ability to switch the primary ignition on and off. Technician B says an ignition module tester proves the module's ability to react to computed timing. Who is right?
    A. A only
    B. B only
    C. Both A and B
    D. Neither A nor B (B.9)
65. Technician A says that when checking a pulsed secondary air injection system, erratic pulses could indicate a cylinder misfire. Technician B says when checking a pulsed secondary air injection system, there should be steady audible pulses. Who is right?
    A. A only
    B. B only
    C. Both A and B
    D. Neither A nor B (D.3.1)

66. Which of the following would most likely cause weak spark at the spark plug wires?
    A. Timing advance out of spec
    B. Leaking secondary insulation
    C. Low resistance in the secondary circuit
    D. High resistance in the primary circuit (A.9, B.5)
67. A voltmeter is connected across a 12-volt battery. With the engine cranking, the voltmeter should not read less than:
    A. 12 volts.
    B. 10.5 volts.
    C. 9.6 volts.
    D. 7.5 volts. (F.1.1, F.2.1)
68. Technician A says a leaking cold start injector could cause a high CO reading at idle. Technician B says that no vacuum to the pressure regulator could cause a high CO reading at idle. Who is right?
    A. A only
    B. B only
    C. Both A and B
    D. Neither A nor B (C.6, C.7)
69. A vacuum leak is suspected for a rough idle concern. Using a four-gas analyzer, Technician A says O_2 will be higher than normal. Technician B says CO will be higher than normal. Who is right?
    A. A only
    B. B only
    C. Both A and B
    D. Neither A nor B (A.10, C.1)
70. A multi-trace oscilloscope can test all of the following EXCEPT:
    A. air:fuel ratio.
    B. manifold absolute pressure sensor.
    C. throttle position sensor.
    D. crank position sensor. (A.9)
71. To check coil available voltage output, the technician should:
    A. disconnect the fuel pump power lead.
    B. disconnect the plug wire at the plug and ground it.
    C. disconnect the coil wire and ground it.
    D. conduct the test using a suitable spark tester that requires 25 kV. (B.5)
72. Technician A says to check for battery drain, install a voltmeter from the positive battery terminal to the negative battery terminal. Technician B says to install an ohmmeter from the negative battery cable to a known good ground. Who is right?
    A. A only
    B. B only
    C. Both A and B
    D. Neither A nor B (F.1.2)
73. A mechanical fuel pump is being tested using a standard vacuum/pressure gauge at the fuel pump inlet. Technician A says this checks the condition of the diaphragm. Technician B says this checks the valve in the pump. Who is right?
    A. A only
    B. B only
    C. Both A and B
    D. Neither A nor B (C.2)

74. Technician A says incorrect valve timing can cause an engine not to start. Technician B says incorrect valve timing may cause a power loss. Who is right?
    A. A only
    B. B only
    C. Both A and B
    D. Neither A nor B (A.12)

75. High current draw and low cranking speed usually indicate:
    A. excessive resistance in the starter circuit.
    B. a defective battery.
    C. a defective starter.
    D. a defective ignition switch. (F.2.1)

76. A scan tool is being used in addition to a thermometer to check thermostat operation. Technician A says both tools are required. Technician B says thermostat operation can also be checked visually by running the engine until it is hot. Who is right?
    A. A only
    B. B only
    C. Both A and B
    D. Neither A nor B (A.13)

77. On an idling vehicle equipped with an AIR system, Technician A says the "mil" will turn on after two minutes. Technician B says that the air helps cool the O_2 sensor. Who is right?
    A. A only
    B. B only
    C. Both A and B
    D. Neither A nor B (D.3.2, D.3.3)

78. Technician A says that some resistance in the power distribution circuits may harm the ignition switch or starter due to low voltage. Technician B says the power to the ignition switch is never fused. Who is right?
    A. A only
    B. B only
    C. Both A and B
    D. Neither A nor B (B.3, E.6)

79. A battery has been on a slow charger and the specific gravity is 1.15. Technician A says the battery is completely charged. Technician B says the electrolyte temperature should not exceed 125°F while recharging the battery. Who is right?
    A. A only
    B. B only
    C. Both A and B
    D. Neither A nor B (F.1.1)

80. A technician connects a jumper wire between the alternator B+ and F terminals during a field circuit and alternator test. Technician A says if this corrects a low voltage reading, the wiring harness from the alternator to the regulator could be faulty. Technician B says this bypasses the voltage regulator. Who is right?
    A. A only
    B. B only
    C. Both A and B
    D. Neither A nor B (F.3.1, F.3.2)

81. During cranking, Technician A says battery voltage should be no more than 5 volts. Technician B says cranking voltage doesn't matter, as long as the engine starts. Who is right?
    A. A only
    B. B only
    C. Both A and B
    D. Neither A nor B (F.1.1, F.2.1)
82. During a cylinder leak down test on a four-cylinder engine, air is heard coming from the #3 spark plug hole as cylinder #4 is being checked. Technician A says that this could be caused by a blown head gasket. Technician B says this could be caused by a cracked engine block. Who is right?
    A. A only
    B. B only
    C. Both A and B
    D. Neither A nor B (A.8)
83. Technician A says to adjust idle mixture on older vehicles, use the lean-drop method. Technician B says to adjust idle mixture on older vehicles, use the CO specification method. Who is right?
    A. A only
    B. B only
    C. Both A and B
    D. Neither A nor B (C.13)
84. An MFI vehicle has poor fuel economy, but runs fine. Technician A says that the fuel pressure regulator could be stuck. Technician B says that the fuel return line could be restricted. Who is right?
    A. A only
    B. B only
    C. Both A and B
    D. Neither A nor B (C.5, C.6)
85. When discussing pickup coil testing, all of the following are true EXCEPT:
    A. each pickup coil should be within the manufacturer's specified resistance value.
    B. a resistance reading below manufacturer's specification indicates an open pickup coil.
    C. a resistance reading above manufacturer's specification indicates an open pickup coil.
    D. an erratic reading while wiggling the pickup coil wires indicates that the pickup coil is intermittent. (B.8)
86. The LEAST likely cause of spark knock is:
    A. EGR valve stuck closed.
    B. fuel quality.
    C. spark plugs.
    D. exhaust gas recirculation (EGR) valve stuck open. (D.2.1)
87. A turbocharged engine lacks power. Technician A says this could be caused by a restricted air intake. Technician B says this could be caused by stochiometric air-fuel mixture. Who is right?
    A. A only
    B. B only
    C. Both A and B
    D. Neither A nor B (C.16)

88. A vehicle is equipped with a vented gas cap. Technician A says if a nonvented cap is installed in this vehicle, the gas tank could collapse. Technician B says that if a nonvented cap is installed, the vehicle could be starved for fuel at high speeds. Who is right?
    A. A only
    B. B only
    C. Both A and B
    D. Neither A nor B (C.4)

89. Technician A says corrosion will add resistance when performing a voltage drop test. Technician B says corrosion on ground wires can increase circuit resistance. Who is right?
    A. A only
    B. B only
    C. Both A and B
    D. Neither A nor B (E.5)

90. Technician A says fuel tank problems require replacement. Technician B says most repairs require the tank to be removed. Who is right?
    A. A only
    B. B only
    C. Both A and B
    D. Neither A nor B (C.4)

91. Technician A says the entry portion of a three-way catalytic converter is used to control carbon monoxide and carbon dioxide. Technician B says that the reduction bed of a three-way catalytic converter is used to control oxides of nitrogen. Who is right?
    A. A only
    B. B only
    C. Both A and B
    D. Neither A nor B (D.3.1)

92. Technician A says corrosion on terminals results in high resistance. Technician B says corrosion can occur where the insulation on wires has been pierced or damaged. Who is right?
    A. A only
    B. B only
    C. Both A and B
    D. Neither A nor B (E.5, E.6, F.2.2)

93. Technician A says that if the alternator output is zero, the field circuit may be open. Technician B says that if the belt tension is OK and the output is low, but not zero, the alternator is defective. Who is right?
    A. A only
    B. B only
    C. Both A and B
    D. Neither A nor B (F.3.1)

94. A port fuel injected engine runs fine at idle, but hesitates under acceleration with no DTCs stored. Technician A says to check for a restricted mass airflow sensor. Technician B says to check for a stuck throttle position sensor. Who is right?
    A. A only
    B. B only
    C. Both A and B
    D. Neither A nor B (E.3, C.1, C.7)

95. A port-injected engine is believed to have inoperative fuel cutoff during deceleration. Which of these should the technician do first?
    A. Listen to injector at various speeds.
    B. Check injector timing on deceleration.
    C. Check base engine timing.
    D. Check fuel pressure. (C.7)

96. The LEAST likely test to be performed with a four-gas emission analyzer is the:
    A. cylinder head gasket leak test.
    B. oxygen sensor wave forming test.
    C. cylinder misfiring test.
    D. inspection and maintenance exhaust emission program analysis. (A.9)

97. A battery is being tested with a carbon pile. Technician A says to maintain the load for thirty seconds. Technician B says to apply a load equal to the cold crank rating of the battery. Who is right?
    A. A only
    B. B only
    C. Both A and B
    D. Neither A nor B (F.1.1)

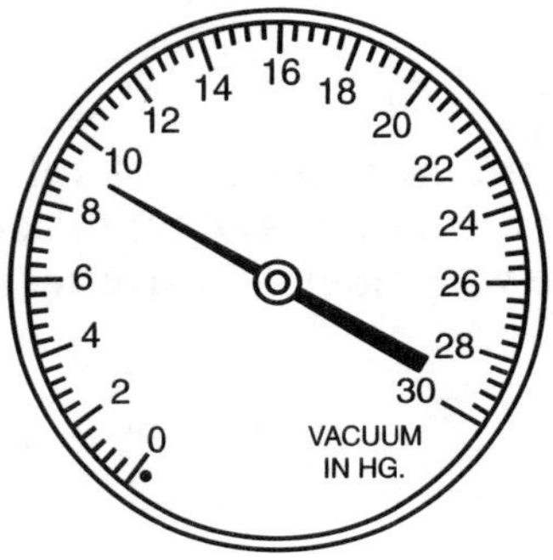

98. The vacuum gauge in the figure indicates low vacuum. Technician A says late ignition timing will cause a low vacuum reading. Technician B says to connect the gauge to a ported vacuum port. Who is right?
    A. A only
    B. B only
    C. Both A and B
    D. Neither A nor B (A.4)

99. All of the following are true of timing the distributor EXCEPT:
    A. the crankshaft timing pointer must be aligned with the timing indicator.
    B. an indicator mark placed before removal aids in reassembly.
    C. on a four-cylinder engine, the distributor can be timed just after TDC on the #3 cylinder.
    D. the ignition rotor must point at the distributor cap terminal for the specified cylinder. (B.4, B.7)

100. A turbocharged engine is experiencing excessive oil consumption and blue smoke from the tailpipe. This could be caused by:
    A. a dirty oil filter.
    B. a sludged center housing.
    C. a dirty air filter.
    D. a broken exhaust pipe. (C.16)

101. An EFI engine has poor acceleration when the throttle is suddenly opened wide. Idle and cruise performances are fine. Technician A says a shorted airflow sensor could cause this. Technician B says faulty throttle position switch contacts could cause this. Who is right?
   A. A only
   B. B only
   C. Both A and B
   D. Neither A nor B (C.1)

102. A vehicle with an electronic ignition fails to start. Technician A says this could be caused by a defective crankshaft sensor connection. Technician B says this could be caused by a defective ignition module. Who is right?
   A. A only
   B. B only
   C. Both A and B
   D. Neither A nor B (C.1)

103. All are inputs for timing control, EXCEPT:
   A. knock sensor.
   B. engine speed (rpm).
   C. power steering load.
   D. engine load. (B.1, B.7, E.8)

104. Technician A says some computer inputs are received from other computers. Technician B says one input might affect other computers. Who is right?
   A. A only
   B. B only
   C. Both A and B
   D. Neither A nor B (E.8)

105. Battery electrolyte specific gravity between cells should not vary by more than:
   A. 0.10.
   B. 0.50.
   C. 0.050.
   D. 0.005. (F.1.1)

106. Technician A says there are multiple ways to access DTCs on some vehicles. Technician B says most technicians use a scan tool for diagnosis. Who is right?
   A. A only
   B. B only
   C. Both A and B
   D. Neither A nor B (E.1)

107. Technician A says to test digital EGR valves a scan tool can be used to cycle it on and off. Technican B says to use a vacuum hand pump to test digital EGR valves. Who is right?
   A. A only
   B. B only
   C. Both A and B
   D. Neither A nor B (D.2.2, D.2.3)

108. Fuel injectors should only be cleaned with:
   A. fuel tank additives.
   B. the fuel pump disconnected.
   C. air pressure.
   D. a brass wire brush. (C.9)

109. Air is escaping from the PCV valve during a cylinder leakage test. Technician A says a blown head gasket is the cause. Technician B says air escaping from the PCV valve is normal. Who is right?
   A. A only
   B. B only
   C. Both A and B
   D. Neither A nor B (A.8)

110. The LEAST likely condition a manifold absolute pressure (MAP) sensor can cause is:
   A. a rich or lean air:fuel ratio.
   B. engine surging.
   C. excess fuel consumption.
   D. excessive idle speeds. (C.1)

111. Low battery or system voltage can cause all of the following, EXCEPT:
   A. increased ignition dwell.
   B. increased injector on-time.
   C. increased idle speed.
   D. increased steering effort. (E.3, F.1.1)

112. Technician A says to use a scan tool to verify coolant temperature sensor input and related DTCs. Technician B says a coolant temperature sensor input is used to help determine loop status. Who is correct?
   A. A only
   B. B only
   C. Both A and B
   D. Neither A nor B (E.2, E.4)

113. Technician A says an analog voltmeter cannot be used to check an O_2 sensor. Technician B says a test light can be used to check an O_2 sensor. Who is right?
   A. A only
   B. B only
   C. Both A and B
   D. Neither A nor B (E.5)

114. All of the following are measured by a four-gas analyzer EXCEPT:
   A. hydrocarbons (HC).
   B. carbon monoxide (CO).
   C. oxides of nitrogen (NO_x).
   D. oxygen (O_2). (A.9)

115. Technician A says you can test for a parasitic battery drain with a battery tester. Technician B says you must connect an ammeter in series to test for parasitic battery drain. Who is right?
   A. A only
   B. B only
   C. Both A and B
   D. Neither A nor B (F.1.2)

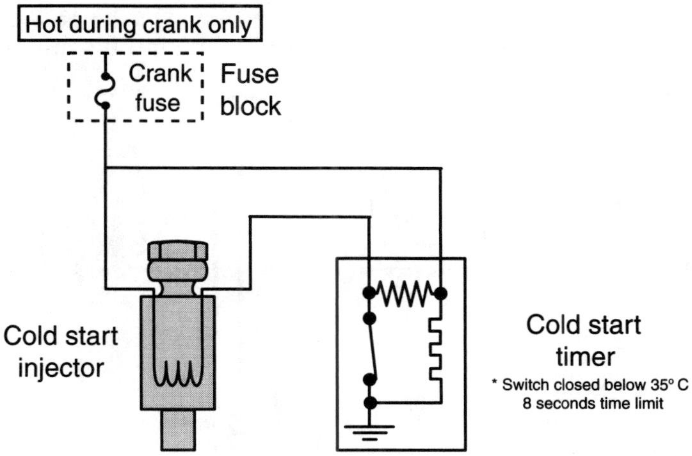

116. Referring to the figure, Technician A says the cold start injector operates during engine cranking only. Technician B says coolant temperature determines how long the cold start injector operates during cranking. Who is right?
    A. A only
    B. B only
    C. Both A and B
    D. Neither A nor B (C.7)

117. All statements are true about adjusting valves EXCEPT:
    A. mechanical lifters are adjusted by rotating the nut on the rocker arm until the specified clearance is achieved.
    B. after hydraulic lifters are pumped up, adjust by rotating the nut until the specified clearance is achieved.
    C. use a feeler gauge for mechanical lifters.
    D. the piston should be at top dead center. (F.1)

118. The customer complains of sluggish performance and poor fuel economy. The vehicle has a DTC for O_2 voltage low/lean exhaust and has high CO exhaust emissions. The O_2 sensor is tested separately and is functioning properly. Technician A says the fuel injectors may be leaking. Technician B says secondary air may be diverted upstream after closed loop status. Who is right?
    A. A only
    B. B only
    C. Both A and B
    D. Neither A nor B (A.10, C.1, C.2, D.3.1, D.3.2, E.1)

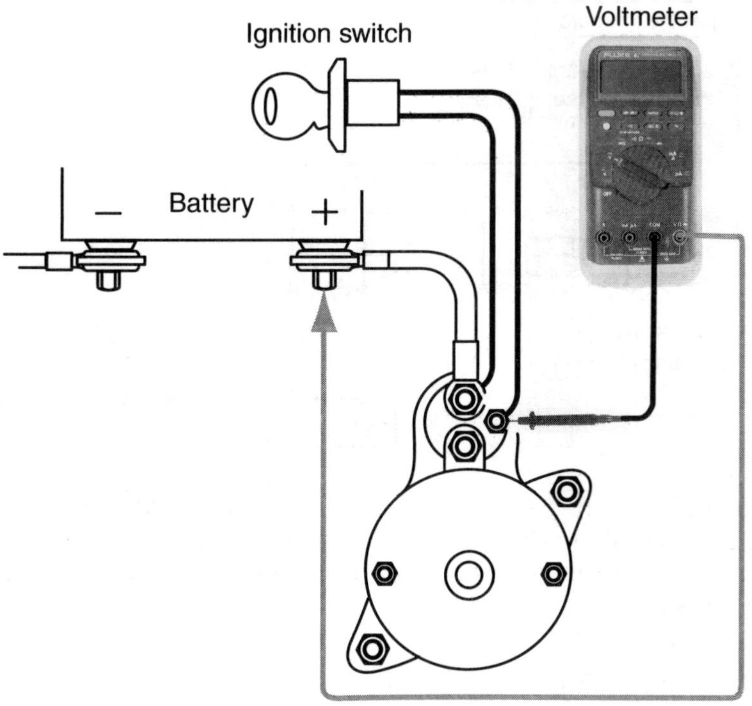

119. The entire starter motor circuit can be checked by connecting the positive voltmeter lead to the positive battery cable at the battery, and connecting the negative voltmeter lead to the solenoid winding terminal on the solenoid, as shown in the figure. Disable the ignition system and place the voltmeter on the lowest scale. Crank the engine. Technician A says that a reading below 2.5 volts is OK. Technician B says that with a reading above 1.5 volts, you will need to check the individual components. Who is right?
    A. A only
    B. B only
    C. Both A and B
    D. Neither A nor B (F.2.2)

120. All of the following apply to OBD-II guidelines, EXCEPT:
    A. use of a standard list of diagnostic trouble codes.
    B. standard communication protocol.
    C. ability to record and store fault conditions when they occur.
    D. turn on the MIL if emission levels exceed four times the standards for that model year vehicle. (E.1, E.2)

121. In discussing a vehicle that has an engine miss under acceleation, and sometimes at cruise speed, but idles smooth, Technician A says a coil with weak available voltage could cause it. Technician B says there may be an engine misfire DTC stored to aid in diagnosis. Who is right?
    A. A only
    B. B only
    C. Both A and B
    D. Neither A nor B (B.1, B.2, B.6)

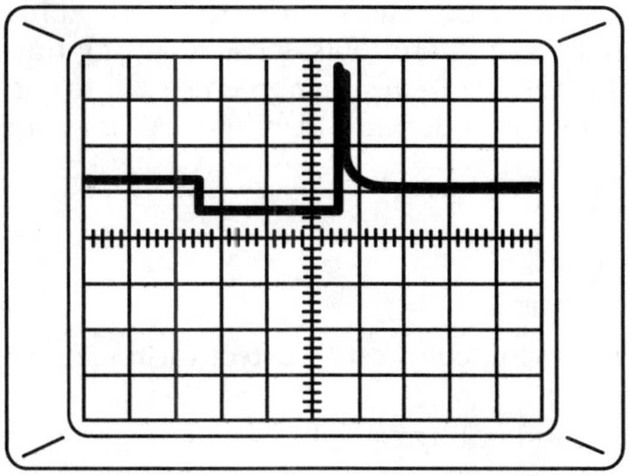

122. Referring to the sample pattern, a vehicle with port fuel injection is running roughly. A lab scope shows each injector waveform to be identical, except for one that has a considerably shorter voltage spike than the others. Which of the following is the most likely cause?
    A. Bad PCM
    B. Open connection at the injector
    C. Shorted injector winding
    D. Low charging system voltage (A.9, C.1)

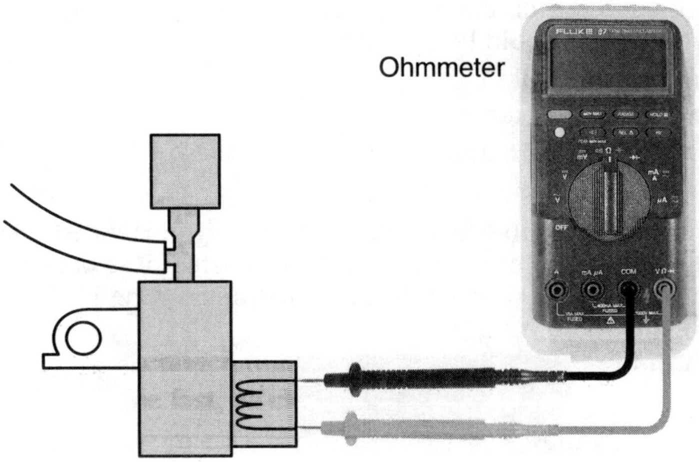

123. A diagnostic trouble code for an EGR problem is retrieved in the figure. A reading of infinite between terminals of the EGR vacuum regulator could mean:
    A. nothing.
    B. the regulator is defective.
    C. the EGR valve is defective.
    D. the MAP sensor is defective. (D.3.2, D.3.3)

124. Technical service bulletin information is:
    A. for dealership technicians only.
    B. essential for information updates or corrections.
    C. published for recall vehicles only.
    D. beneficial only when diagnosing computer controlled systems. (A.2)

125. Technician A says when high resistance is found in a circuit, check for burned wires, connector ring terminals, loose retaining nuts, or other wire and connector concerns. Technician B says that greenish white corrosion can happen at any point where the wire insulation has been pierced or opened in any way. Who is right?
   A. A only
   B. B only
   C. Both A and B
   D. Neither A nor B (E.5)

126. All of the following checks are correct for testing the coolant sensor and/or circuitry EXCEPT:
   A. resistance and voltage checks.
   B. DTCs and scan data.
   C. thermometer, heated water, and resistance check.
   D. diode check. (E.2, E.4)

127. Technician A says that the first step of any diagnostic procedure is to check for diagnostic trouble codes (DTCs). Technician B says that the customer complaint should be verified before performing any diagnostic procedures. Who is right?
   A. A only
   B. B only
   C. Both A and B
   D. Neither A nor B (A.1)

128. A primary ignition circuit on a vehicle checks good, but there is no spark from the coil wire. This could be caused by:
   A. a defective coil.
   B. a grounded rotor.
   C. an overheated transistor.
   D. an open diode. (B.6)

129. A multiport fuel injected vehicle has a rough extended idle. Technician A says this could be caused by a cracked or disconnected hose between the fuel tank and the EVAP canister. Technician B says a malfunctioning EVAP system can cause idle problems. Who is right?
   A. A only
   B. B only
   C. Both A and B
   D. Neither A nor B (D.4.1, D.4.3)

130. Technician A says the oxygen sensor can be removed and tested under different temperatures. Technician B says you can use a scan tool to check for oxygen sensor codes and operation. Who is right?
   A. A only
   B. B only
   C. Both A and B
   D. Neither A nor B (C.2, E.2)

131. Many AIR system pumps have a serviceable pulley. Technician A says many also have a replaceable filter. Technician B says that when replacing the pulley, you must also replace the drive belt. Who is right?
   A. A only
   B. B only
   C. Both A and B
   D. Neither A nor B (D.3.3)

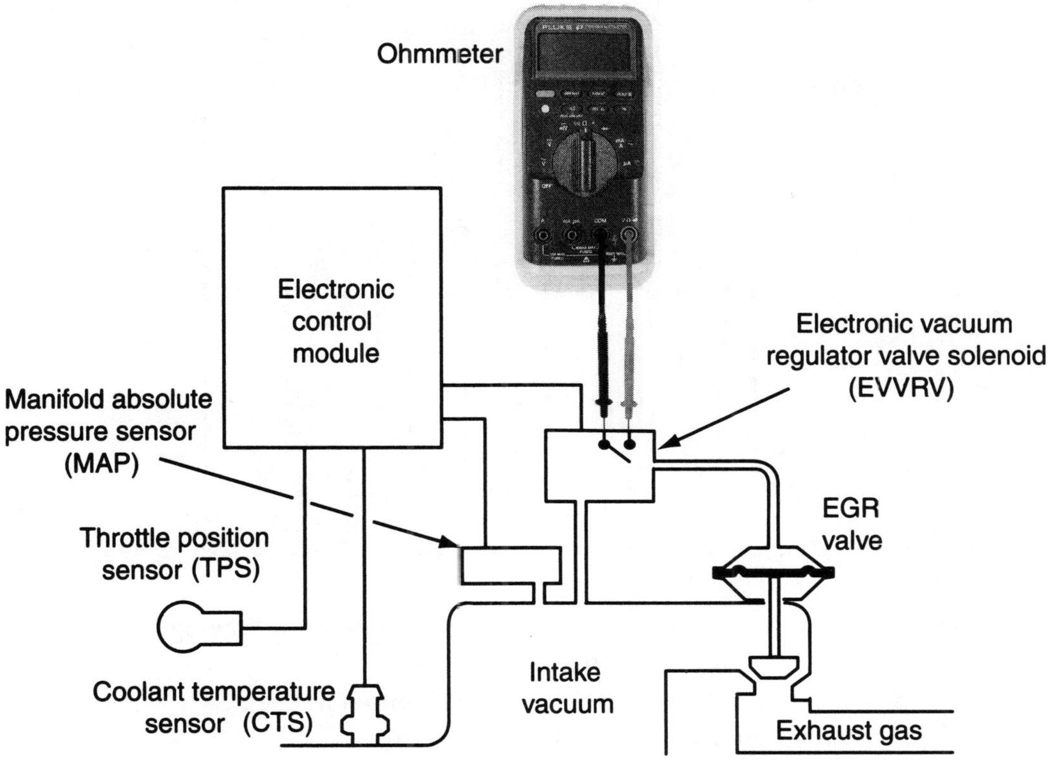

132. In the figure, an infinite resistance reading between the terminals of the EGR vacuum regulator (EVR) could mean any of the following EXCEPT:
    A. the vacuum regulator is normally open.
    B. the exhaust gas recirculation (EGR) valve is defective.
    C. nothing.
    D. prove that the regulator is defective. (D.2.1, D.2.3)

133. While discussing electronic ignition service and diagnosis, Technician A says the camshaft sensor may be rotated to adjust the base ignition timing. Technician B says on some systems, the crankshaft sensor may be moved to obtain proper clearance between the sensor and interrupter blades. Who is right?
    A. A only
    B. B only
    C. Both A and B
    D. Neither A nor B (B.8)

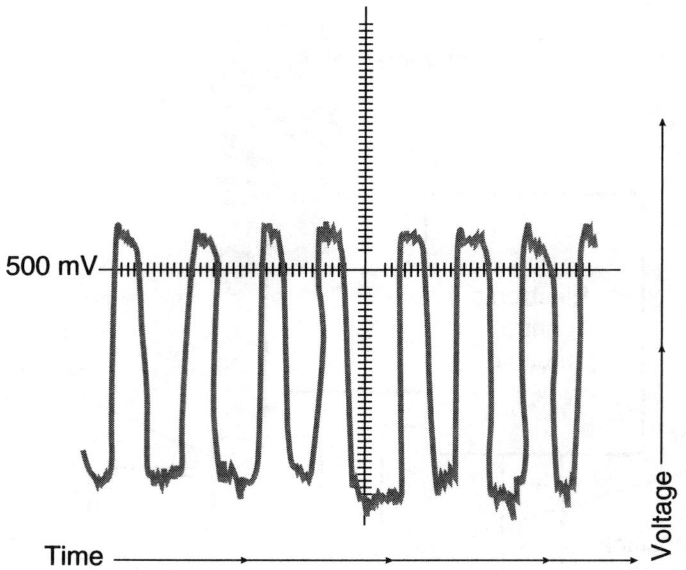

134. Based on the O_2 sensor wave form shown, all the following are true EXCEPT:
   A. this represents a lean biased condition.
   B. a diagnostic trouble code may be recorded in the PCM.
   C. this represents a rich biased condition.
   D. the O_2 sensor is functioning. (C.2, E.2, E.4)

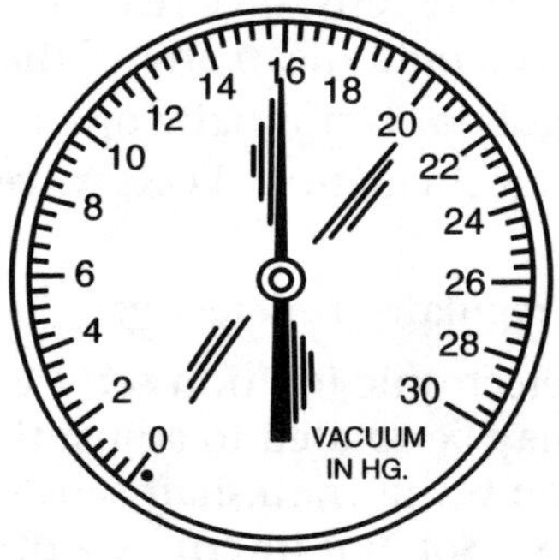

135. The vacuum gauge shows a fluctuating motion from 15 to 21 in. Hg at idle. Technician A says this could be caused by a loose exhaust manifold. Technician B says this could be caused by a burned valve. Who is right?
   A. A only
   B. B only
   C. Both A and B
   D. Neither A nor B (A.5)

136. The ignition module uses a digital signal from the PCM for:
   A. timing advance control.
   B. #1 TDC signal.
   C. rpm input.
   D. fuel economy information. (B.7)

137. Technician A says that worn valve train components usually produce an identifiable noise. Technician B says an engine noise diagnosis should be performed before doing engine repair work. Who is right?
A. A only
B. B only
C. Both A and B
D. Neither A nor B (A.3)

138. A faulty fuel pump is suspected. Which of these steps should the technician take first?
A. Perform fuel pump pressure and volume tests.
B. Check for fuel pump diagnostic trouble codes (DTC).
C. Check the fuel filter.
D. Check the fuel lines. (C.3)

139. Technician A says that some computer inputs are received from another computer. Technician B says that a communication problem that sets a diagnostic trouble code (DTC) in only one computer could be faulty circuitry. Who is right?
A. A only
B. B only
C. Both A and B
D. Neither A nor B (E.8)

140. While discussing adaptive fuel control, Technician A says the PCM will increase the fuel injector pulse width if there is excess oxygen in the exhaust. Technician B says if there is a lean condition, the SFT will show a minus value on the scan tool. Who is right?
A. A only
B. B only
C. Both A and B
D. Neither A nor B (E.4)

141. Technician A says prior to adjusting the throttle plates on a multiport injected throttle body, the idle air control valve must be fully extended. Technician B says the throttle position sensor needs to be readjusted on some vehicles after the throttle plate angle adjustment is complete. Who is right?
A. A only
B. B only
C. Both A and B
D. Neither A nor B (C.8, E.4)

142. While discussing MAP sensors, Technician A says a MAP sensor should be able to hold vacuum during a test. Technician B says some MAP sensors are a digital type. Who is right?
A. A only
B. B only
C. Both A and B
D. Neither A nor B (C.13, E.2, E.3, E.4)

143. A faulty brake switch may affect all the following, EXCEPT:
A. idle speed.
B. TCC engagement.
C. EGR function.
D. throttle position sensor signal voltage. (E.8)

7 Appendices

Answers to the Test Questions for the Sample Test Section 5

1.	B	21.	B	41.	C	61.	A
2.	A	22.	D	42.	C	62.	B
3.	A	23.	A	43.	C	63.	C
4.	D	24.	C	44.	C	64.	C
5.	B	25.	C	45.	C	65.	A
6.	C	26.	B	46.	C	66.	A
7.	B	27.	B	47.	D	67.	B
8.	A	28.	A	48.	C	68.	C
9.	C	29.	D	49.	B	69.	B
10.	A	30.	C	50.	D	70.	C
11.	A	31.	D	51.	C	71.	B
12.	D	32.	B	52.	C	72.	C
13.	A	33.	D	53.	C	73.	D
14.	B	34.	C	54.	B	74.	A
15.	D	35.	B	55.	B	75.	D
16.	C	36.	B	56.	B	76.	C
17.	D	37.	A	57.	C	77.	C
18.	D	38.	A	58.	D		
19.	C	39.	C	59.	A		
20.	B	40.	C	60.	A		

Explanations to the Answers for the Sample Test Section 5

Question #1
Answer A is wrong. The level of the electrolyte is critical to battery performance.
Answer B is correct. Signs of leakage indicate loss of electrolyte. The battery should be removed from service.
Answer C is wrong.
Answer D is wrong.

Question #2
Answer A is correct.
Answer B is wrong. Road damage often causes fuel tank damage.
Answer C is wrong. Defective seams can be a cause of fuel leaks.
Answer D is wrong. Corrosion, after it is through the tank, will cause a fuel leak.

Question #3
Answer A is correct. Only bend in gradual curves.
Answer B is wrong. Nylon tubing cannot make sharp bends or turns. This will cause permanent damage and reduce fuel flow.
Answer C is wrong.
Answer D is wrong.

Question #4
Answer A is wrong. Improper cooling affects the life of the bearings.
Answer B is wrong. Not changing the oil frequently causes sludge buildup.
Answer C is wrong. A dirty air cleaner will contaminate the air intake with dust and dirt, which shorten the life of the turbocharger.
Answer D is correct.

Question #5
Answer A is wrong. Starter draw is checked in amps.
Answer B is correct. Any "live" circuit will cause amperage flow.
Answer C is wrong. Voltage is measured in volts.
Answer D is wrong. Voltage drops are measured in volts, not milliamps.

Question #6
Answer A is wrong because B is also correct.
Answer B is wrong because A is also correct.
Answer C is correct. The pressure test is done to prove pump operation, thus confirming power, ground, and pump action. Specifications are published for fuel pressure and volume.
Answer D is wrong.

Question #7
Answer A is wrong. Not all valve adjustments require the engine to be cold. Some manufacturers specify a warm engine.
Answer B is correct. Many adjustment procedures require the piston at top dead center on the compression stroke as the intake and exhaust valves are closed.
Answer C is wrong because A is wrong.
Answer D is wrong because B is correct.

Question #8
Answer A is correct. Hotter temperatures result in more resistance, thus causing the fan to move more air across the radiator.
Answer B is wrong. It has less resistance cold, when fan-blown air is not needed, to improve performance and fuel economy.
Answer C is wrong. It should have movement.
Answer D is wrong. The clutch should have resistance.

Question #9
Answer A is wrong. 0 percent indicates that there is no leakage.
Answer B is wrong. Air escaping from the crankcase could indicate problems with the rings.
Answer C is correct. A reading of 100 percent indicates a major, total cylinder leak, either due to an incorrect crank position (valve open) or internal damage. A reading of up to 20 percent is considered normal.
Answer D is wrong. Air escaping from the exhaust indicates a bad exhaust valve.

Question #10
Answer A is correct. Most engines will require spark delivery to occur several degrees before the piston reaches top dead center.
Answer B is wrong. Having a clean mating surface is critical to proper sealing.
Answer C is wrong. A true mating surface is important to proper sealing.
Answer D is wrong. Exhaust gas can leak through cracks easily.

Question #11
Answer A is correct. Most engines will require spark delivery to occur several degrees before the piston reaches top dead center.
Answer B is wrong. The engine should be timed with the engine on the compression stroke.
Answer C is wrong.
Answer D is wrong.

Question #12
Answer A is wrong. Since the valve is located downstream from the throttle plates, there will always be vacuum at the PCV valve.
Answer B is wrong. There should be a noise when the valve is shaken, proving the plunger is not stuck.
Answer C is wrong.
Answer D is correct.

Question #13
Answer A is correct. Sealing of mating surfaces has nothing to do with module replacement.
Answer B is wrong. Procedures do vary between applications.
Answer C is wrong. Dielectric silicone is used for heat dissipation.
Answer D is wrong. Damage may occur if the module cannot dissipate heat.

Question #14
Answer A is wrong. Idle speed (throttle plate angle) and ignition timing should always be adjusted prior to final mixture adjustment.
Answer B is correct. All related engine systems, connections, vacuum hoses, and components should be inspected before any final adjustments are made.
Answer C is wrong.
Answer D is wrong.

Question #15
Answer A is wrong. The battery should be charged at one-tenth of the ampere-hour rating.
Answer B is wrong. The battery should be charged until specific gravity is above 1.250.
Answer C is wrong.
Answer D is correct.

Question #16
Answer A is wrong because B is also correct.
Answer B is wrong because A is also correct.
Answer C is correct. Some vehicles will use one sensor input to both engine and body controllers. Improper signals can affect multiple systems.
Answer D is wrong.

Question #17
Answer A is wrong. If the pickup coil leads are moved, there should be no erratic reading on the ohmmeter. If there is, look for damaged pickup coil wires.
Answer B is wrong. An infinite reading indicates an open circuit.
Answer C is wrong.
Answer D is correct.

Question #18
Answer A is wrong. The ignition module does not need an rpm signal.
Answer B is wrong. There is no such thing as hall effect timing.
Answer C is wrong. The ignition module does not discriminate between cylinders.
Answer D is correct. The signal from the PCM is the result of computed inputs for proper timing advance. This is the signal the ignition module uses to fire the coil at the proper time.

Question #19
Answer A is wrong because B is also correct.
Answer B is wrong because A is also correct.
Answer C is correct. Any vacuum leak will change the air-fuel mixture, resulting in loss of performance and fuel economy. Controlled use of propane gas can help confirm and locate the source of vacuum leaks.
Answer D is wrong.

Question #20
Answer A is wrong. A timing belt failure would cause the engine to quit running.
Answer B is correct. Incorrect valve timing effects volumetric efficiency.
Answer C is wrong.
Answer D is wrong.

Question #21
Answer A is wrong. Fuel cut strategy involves more than cutting off the fuel injectors when a speed is reached.
Answer B is correct. Reducing fuel on deceleration or while coasting helps reduce emissions and reduces the workload on the catalytic converter. Some vehicles also have fuel shut off programmed for over-revving protection when exceeding a specific rpm.
Answer C is wrong.
Answer D is wrong.

Question #22
Answer A is wrong. When replacing a PROM, always ground yourself to the vehicle, to reduce static discharge.
Answer B is wrong. Being grounded will not erase the PROM.
Answer C is wrong.
Answer D is correct.

Question #23
Answer A is correct. Incorrect inputs can result in performance or emission problems.
Answer B is wrong. A faulty TPS sensor does not always cause a fault code.
Answer C is wrong.
Answer D is wrong.

Question #24
Answer A is wrong.
Answer B is wrong.
Answer C is correct. A restricted canister filter can prevent fresh airflow through the canister, causing ineffective purging. It should be checked or replaced at recommended service intervals. Gas caps can be checked for pressure and vacuum holds. There is equipment and specifications available for this test.
Answer D is wrong.

Question #25
Answer A is wrong. Researching service bulletins saves diagnosis time.
Answer B is wrong. Knowing the year, model, and make is critical.
Answer C is correct. Specifications are found in the service manual. Changes to specifications may be found in bulletins.
Answer D is wrong. Midyear production changes are helpful to know if there are design changes that can affect the diagnosis.

Question #26
Answer A is wrong. A rough idle would indicate a malfunction in the evaporative emission system.
Answer B is correct. A malfunction would usually result in increased emissions.
Answer C is wrong. A MIL lamp is a good indication that there is a problem.
Answer D is wrong. Evaporative emissions leaks are indicated by a raw fuel odor.

Question #27
Answer A is wrong. The evaporative emission system recovers HC emissions from the fuel tank.
Answer B is correct. The canister traps vapor and purges while driving. The vapors are carefully metered and are eliminated through the combustion process.
Answer C is wrong.
Answer D is wrong.

Question #28
Answer A is correct. The AIR pump provides additional air to the exhaust and catalytic converter. This helps protect the converter and continues the burning of exhaust gas. Disabling the pump, or restricting airflow, will cause a reduction of O_2 at least by 2 percent. This can be verified with a four-gas exhaust analyzer.
Answer B is wrong. The O_2 readings will change when the AIR system is disconnected.
Answer C is wrong.
Answer D is wrong.

Question #29
Answer A is wrong. An AIRB solenoid could cause this problem.
Answer B is wrong. An air supply line could cause this problem.
Answer C is wrong. The valve itself could be the problem.
Answer D is correct. A one-way check valve is not a component of this part of the system.

Question #30
Answer A is wrong.
Answer B is wrong.
Answer C is correct. The valve diverts the air to the atmosphere under certain conditions. If the valve fails, the flow may not go to the exhaust, thus increasing emissions.
Answer D is wrong.

Question #31
Answer A is wrong. An infinite reading indicates only an open winding.
Answer B is wrong. An infinite reading does not indicate a shorted winding.
Answer C is wrong.
Answer D is correct.

Question #32
Answer A is wrong. A plugged PCV valve would be considered part of the PCV system.
Answer B is correct. Excessive blowby suggests an internal engine problem.
Answer C is wrong. Hoses are part of the PCV system.
Answer D is wrong. Oil in the air cleaner housing can be an indication of the PCV system not functioning properly.

Question #33
Answer A is wrong. A damaged turbo bearing will limit the impeller speed.
Answer B is wrong. An open wastegate will not allow the impeller to build up pressure.
Answer C is wrong. A leak limits the maximum pressure in the cylinder.
Answer D is correct. Coolant temperature sensors are not part of the turbocharger.

Question #34
Answer A is wrong.
Answer B is wrong.
Answer C is correct. Exhaust gases that are restricted will affect volumetric efficiency, interfering with the new incoming air-fuel charge.
Answer D is wrong.

Question #35
Answer A is wrong. An open PCV valve will lean the mixture, not richen it.
Answer B is correct. Crankcase pressure builds, causing blowby.
Answer C is wrong.
Answer D is wrong.

Question #36
Answer A is wrong.
Answer B is correct. Verification includes re-creating the conditions in which the complaint occurred, in an attempt to duplicate it. Stored DTCs are a result of a problem, not necessarily the reason or specific definition of the fault.
Answer C is wrong.
Answer D is wrong.

Question #37
Answer A is correct. This type of thermister is known as a negative temperature coefficient (NTC) type.
Answer B is wrong. As the water temperature increases, the sensor value decreases.
Answer C is wrong.
Answer D is wrong.

Question #38
Answer A is correct.
Answer B is wrong. Some tests of the EGR valve should be done at idle engine speeds.
Answer C is wrong.
Answer D is wrong.

Question #39
Answer A is wrong because B is also correct
Answer B is wrong because A is also correct
Answer C is correct. An engine misfire causes a disruption of airflow due to a change in exhaust flow rate.
Answer D is wrong.

Question #40
Answer A is wrong.
Answer B is wrong.
Answer C is correct. The lower the vacuum on the full pressure regulator, the higher the fuel pressure will be. Obviously, there is no vacuum while the engine is not running.
Answer D is wrong.

Question #41
Answer A is wrong.
Answer B is wrong.
Answer C is correct. The cooling system may have more than one leak. The cap must be tested to confirm it will hold pressure. There is a specification for the system pressure.
Answer D is wrong.

Question #42
Answer A is wrong. Sticking valves would not make the vacuum gauge drop.
Answer B is wrong. Advanced ignition timing would not cause vacuum to decrease.
Answer C is correct. Exhaust restrictions cause pressure to build due to the presence of gases in the combustion chamber.
Answer D is wrong. Fuel mixture would not cause changes in vacuum readings.

Question #43
Answer A is wrong.
Answer B is wrong.
Answer C is correct. MAP sensors indicate engine load. By testing under varying vacuum amounts, the complete range of values can be verified.
Answer D is wrong.

Question #44
Answer A is wrong because B is also correct.
Answer B is wrong because A is also correct.
Answer C is correct. Either condition results in added circuit load, causing excessive voltage drop.
Answer D is wrong.

Question #45
Answer A is wrong. The thermometer should be slightly higher than the thermostat temperature.
Answer B is wrong. It will not be more than a few degrees higher than the thermostat temperature.
Answer C is correct. The temperature should have stabilized close to normal operating temperature.
Answer D is wrong. The temperature should always be slightly higher than the thermostat temperature.

Question #46
Answer A is wrong because B is also correct.
Answer B is wrong because A is also correct.
Answer C is correct. An exhaust restriction prohibits the new air-fuel charge to enter the combustion chamber, resulting in power loss.
Answer D is wrong.

Question #47
Answer A is wrong. Diagnostic trouble codes (DTCs) do not indicate which component to replace.
Answer B is wrong. There is not an emissions reminder lamp that illuminates when an emissions fault occurs.
Answer C is wrong.
Answer D is correct.

Question #48
Answer A is wrong. Hesitation is a sign of a bad accelerator pump.
Answer B is wrong. Stumbles on acceleration are an indication of a faulty accelerator pump.
Answer C is correct. The accelerator pump circuit does not function during steady cruise speeds.
Answer D is wrong. An excessive accelerator pump stroke can cause excessive fuel consumption.

Question #49
Answer A is wrong. An analog voltmeter cannot switch fast enough.
Answer B is correct. The digital voltmeter is a common standard for all automotive computer system circuitry.
Answer C is wrong.
Answer D is wrong.

Question #50
Answer A is wrong. A flickering test lamp indicates that the module is working properly.
Answer B is wrong. A flickering test lamp indicates the pick-up coil is working properly.
Answer C is wrong.
Answer D is correct. If the test lamp flickers, primary switching is occurring, indicating the pickup and module are functioning.

Question #51
Answer A is wrong.
Answer B is wrong.
Answer C is correct. If cylinders located next to each other both indicate a problem, inspect what is common to both. In most cases, it will be a head gasket. If a problem exists in just one cylinder, look for items that control airflow and pressure for that cylinder.
Answer D is wrong.

Question #52
Answer A is wrong. Checking the spray pattern is not the easiest test to perform.
Answer B is wrong. There is only one cold start injector; therefore, it cannot be compared to another.
Answer C is correct. This, in most cases, is the easiest test to do.
Answer D is wrong. It is easier to perform a resistance value check.

Question #53
Answer A is wrong.
Answer B is wrong.
Answer C is correct. Any added resistance to a connection or circuit will cause an increase in voltage drop. This will cause problems in either power or ground circuits.
Answer D is wrong.

Question #54
Answer A is wrong. Defective spark plugs can be diagnosed with a cylinder balance test.
Answer B is correct. Timing is common to all cylinders.
Answer C is wrong. Faulty ignition wires can be determined with cylinder power balance tests.
Answer D is wrong. Burned valves can be diagnosed with cylinder power balance tests.

Question #55
Answer A is wrong. Damaged carbon ignition wires cause high resistance.
Answer B is correct. The ignition module is common to all cylinders and is part of the primary ignition circuit.
Answer C is wrong. Corroded spark plug wire ends will cause high resistance.
Answer D is wrong. An excessive rotor air gap causes high resistance.

Question #56
Answer A is wrong. Head gaskets can be tested with an emission analyzer.
Answer B is correct. This test is done with lab scope or graphing multi-meter.
Answer C is wrong. Misfiring cylinders will raise HC emissions.
Answer D is wrong. Emission tests are performed at regular maintenance intervals.

Question #57
Answer A is wrong.
Answer B is wrong.
Answer C is correct. Glazed, cracked, or deteriorated belts can make noise. The proper fix is to replace the belt. In any case, the proper belt tension adjustment must be made.
Answer D is wrong.

Question #58
Answer A is wrong. The ignition has to be disabled so the engine won't start.
Answer B is wrong. There should be no electrical loads other than the starter.
Answer C is wrong. The doors should be closed, so the interior lamps are out.
Answer D is correct.

Question #59
Answer A is correct. The use of a lab scope or DMM can be used to monitor O_2 switching.
Answer B is wrong. There are other ways to test fuel injectors, including resistance and injector wave form analysis.
Answer C is wrong.
Answer D is wrong.

Question #60
Answer A is correct. Air routing is controlled downstream from the pump itself.
Answer B is wrong. Air injection pump air filters need periodic maintenance.
Answer C is wrong. A bent pulley can cause the air pump to malfunction.
Answer D is wrong. Worn pump shafts often cause malfunctions within the system.

Question #61
Answer A is correct. On vehicles with adjustable timing, specifications and timing setup procedures are on the under hood label.
Answer B is wrong. Timing lights must be connected to only one specified plug wire (usually number one).
Answer C is wrong.
Answer D is wrong.

Question #62
Answer A is wrong. Jumping terminals at the DLC is an approved method for retrieving a DTC.
Answer B is correct. Cycling the lamp does nothing to retrieve codes.
Answer C is wrong. With Chrysler vehicles, the ignition key can be cycled to retrieve codes.
Answer D is wrong. A scan tool always works to retrieve codes.

Question #63
Answer A is wrong because B is also correct.
Answer B is wrong because A is also correct.
Answer C is correct. A stuck EGR valve causes a vacuum leak, while an EGR system that does not operate causes high combustion chamber temperatures, resulting in detonation and NO_x output.
Answer D is wrong.

Question #64
Answer A is wrong.
Answer B is wrong.
Answer C is correct. The higher the resistance, the higher the voltage drop. Voltage drop is defined as voltage expended in pushing a current through a resistance.
Answer D is wrong.

Question #65
Answer A is correct. Damaged filters will not trap dirt, allowing it to pass to the engine.
Answer B is wrong. A restriction in the air inlet system can cause a rich mixture, affecting fuel economy.
Answer C is wrong.
Answer D is wrong.

Question #66
Answer A is correct. If the shaft has too much play, it will allow the wheel to come in contact with the housing, as the unit is designed with close tolerances.
Answer B is wrong. An air leak in the intake cannot cause housing damage.
Answer C is wrong.
Answer D is wrong.

Question #67
Answer A is wrong. This is not a good way to test modules.
Answer B is correct. This method, and others, is used to test modules, eliminating "guess-work" that can result in unnecessary and costly parts changing.
Answer C is wrong.
Answer D is wrong.

Question #68
Answer A is wrong.
Answer B is wrong.
Answer C is correct. Deposits can build up in passageways in the throttle body, thus requiring removal. Any build up on the body or plates will result in restricted or disturbed airflow, causing performance problems.
Answer D is wrong.

Question #69
Answer A is wrong. Restricted passages are always a problem in the exhaust.
Answer B is correct.
Answer C is wrong.
Answer D is wrong.

Question #70
Answer A is wrong.
Answer B is wrong.
Answer C is correct. The belt condition and tension is important to drive the alternator for proper output. If the output is low, the alternator may be the problem. Slip rings and brushes are common wear items in many alternators.
Answer D is wrong.

Question #71
Answer A is wrong. If the voltage drop is 3.5 volts, it is considered to be excessive.
Answer B is correct.
Answer C is wrong.
Answer D is wrong.

Question #72
Answer A is wrong because injector cleaning requires a cleaning solution.
Answer B is wrong because the adaptive memory will have to be reset.
Answer C is correct. The fuel pump must not be activated during the cleaning procedure.
Answer D is wrong because the return line must be restricted to avoid cleaning solution returning to the tank.

Question #73
Answer A is wrong. The fuel pressure gauge connecting point on TBI systems varies.
Answer B is wrong. A plugged filter cannot cause high fuel pressure.
Answer C is wrong.
Answer D is correct.

Question #74
Answer A is correct.
Answer B is wrong. Carbon piles are variable resistors by design, and can be adjusted to control the amount of load applied.
Answer C is wrong.
Answer D is wrong.

Question #75
Answer A is wrong. Black colored smoke is an indication of oil consumption.
Answer B is wrong. Gray colored smoke indicates coolant is being burned; blue is oil.
Answer C is wrong.
Answer D is correct.

Question #76
Answer A is wrong.
Answer B is wrong.
Answer C is correct. A stethoscope is an excellent pinpoint tool to determining the source of an engine noise. In some instances, the noise needs to be duplicated under certain conditions, such as at a specific temperature or load, or rpm.
Answer D is wrong.

Question #77
Answer A is wrong.
Answer B is wrong.
Answer C is correct. The coil has two circuits, a primary and secondary. Both need to be tested for resistance. A dynamic output test using an oscilloscope will show specific voltages.
Answer D is wrong.

Answers to the Test Questions for the Additional Test Questions Section 6

1. B
2. A
3. C
4. B
5. D
6. C
7. C
8. C
9. D
10. A
11. A
12. B
13. D
14. B
15. A
16. C
17. D
18. A
19. B
20. A
21. A
22. D
23. C
24. B
25. C
26. D
27. C
28. B
29. A
30. A
31. D
32. D
33. A
34. D
35. B
36. C
37. A
38. C
39. C
40. A
41. A
42. C
43. B
44. A
45. D
46. C
47. C
48. B
49. D
50. C
51. D
52. C
53. C
54. D
55. B
56. A
57. D
58. D
59. D
60. C
61. D
62. A
63. C
64. A
65. C
66. D
67. C
68. C
69. D
70. A
71. D
72. D
73. C
74. C
75. C
76. D
77. D
78. D
79. B
80. C
81. D
82. C
83. C
84. C
85. B
86. D
87. A
88. C
89. C
90. B
91. B
92. C
93. A
94. C
95. A
96. B
97. D
98. A
99. C
100. B
101. B
102. C
103. C
104. C
105. C
106. C
107. A
108. B
109. D
110. D
111. D
112. C
113. A
114. C
115. B
116. C
117. B
118. B
119. B
120. D
121. C
122. C
123. B
124. B
125. C
126. D
127. B
128. A
129. B
130. C
131. A
132. D
133. B
134. C
135. B
136. A
137. C
138. A
139. C
140. A
141. C
142. C
143. D

Explanations to the Answers for the Additional Test Questions Section 6

Question #1
Answer A is wrong. A squeal is an indication of a loose belt rather than a growl noise that bearings make.
Answer B is correct.
Answer C is wrong.
Answer D is wrong.

Question #2
Answer A is correct. The diverter valve prohibits AIR pump air from entering the exhaust on deceleration. This prevents the continuation of combustion in the exhaust (backfire).
Answer B is wrong. AIR systems do not affect fuel mixture.
Answer C is wrong. Air is not diverted into the passenger compartment.
Answer D is wrong. The AIR system has nothing to do with the A/C system.

Question #3
Answer A is wrong.
Answer B is wrong.
Answer C is correct. Most vehicles will display DTC via a scan tool. Some will display codes by way of pulsing an analog voltmeter connected into the vehicle's diagnostic connector.
Answer D is wrong.

Question #4
Answer A is wrong. Infinity indicates an open circuit. Any resistance in the circuit will display a reading.
Answer B is correct.
Answer C is wrong.
Answer D is wrong.

Question #5
Answer A is wrong. A stethoscope cannot be used to locate a plugged PCV system.
Answer B is wrong. A vacuum gauge cannot be used to diagnose a PCV system.
Answer C is wrong.
Answer D is correct.

Question #6
Answer A is wrong because B is also correct.
Answer B is wrong because A is also correct.
Answer C is correct. Electrical components must be removed before soaking the throttle body in solution.
Answer D is wrong.

Question #7
Answer A is wrong.
Answer B is wrong.
Answer C is correct. The dielectric silicone helps transfer heat created in the module to the heat sink surface. Lack of the compound will allow too much heat buildup and can potentially destroy the component.
Answer D is wrong.

Question #8
Answer A is wrong.
Answer B is wrong.
Answer C is correct. The valve will open late and close early, reducing flow and overlap.
Answer D is wrong.

Question #9
Answer A is wrong. Worn valves could cause low compression.
Answer B is wrong. Worn piston rings can cause low compression.
Answer C is wrong. A blown head gasket can cause low compression.
Answer D is correct.

Question #10
Answer A is correct.
Answer B is wrong. There are no tables from tool manufacturers because all batteries have a load test standard.
Answer C is wrong.
Answer D is wrong.

Question #11
Answer A is correct. This is due to the stopping and starting movement as referenced at top dead center.
Answer B is wrong. A main bearing noise would not have a double knock.
Answer C is wrong.
Answer D is wrong.

Question #12
Answer A is wrong. Carburetor cleaner creates a mess and is a fire hazard.
Answer B is correct.
Answer C is wrong. Water can short out electrical components.
Answer D is wrong. Small vacuum leaks cannot be heard.

Question #13
Answer A is wrong. Worn piston rings can cause blue smoke from the exhaust.
Answer B is wrong. A damaged cylinder head can cause oil smoke from the exhaust.
Answer C is wrong. Worn valve seals can leak oil into the cylinder which causes blue smoke.
Answer D is correct.

Question #14
Answer A is wrong. The choke would not be applied when the engine is warm.
Answer B is correct. A worn accelerator pump or misadjusted accelerator pump linkage can cause a hesitation.
Answer C is wrong. A fuel filter that is faulty would cause problems other than a tip-in hesitation.
Answer D is wrong. A fuel return line would cause problems all the time.

Question #15
Answer A is correct. Sulfur odor is usually from a rich condition.
Answer B is wrong. Lack of power is a symptom of low fuel pressure.
Answer C is wrong. Engine surging is a symptom of low fuel pressure.
Answer D is wrong. Limited top speed is a symptom of low fuel pressure.

Question #16
Answer A is wrong. The battery is sulfated.
Answer B is wrong. The battery is sulfated.
Answer C is correct.
Answer D is wrong. The battery is sulfated.

Question #17
Answer A is wrong. A straightedge should be used with feeler gauges.
Answer B is wrong. A straightedge should be used with feeler gauges.
Answer C is wrong.
Answer D is correct.

Question #18
Answer A is correct.
Answer B is wrong. A dirty air filter will not cause a no-start condition.
Answer C is wrong. Poor fuel economy would be noticed before engine wear.
Answer D is wrong. Excessive engine wear would have to occur before the engine would use an excessive amount of oil.

Question #19
Answer A is wrong. Shaking the valve is not a way to see if it is clogged.
Answer B is correct.
Answer C is wrong. Using a vacuum to test the PCV system is not a good practice.
Answer D is wrong. Signs of oil residue are not an indication that the PCV system is not functioning properly.

Question #20
Answer A is correct. The crankshaft sensor provides signal voltage to the DIS module to begin primary ignition switching.
Answer B is wrong.
Answer C is wrong. The camshaft sensor is not used with this type of system.
Answer D is wrong.

Question #21
Answer A is correct. On some systems, this is commonly referred to as "bypass mode" where the engine is at initial or base timing. On engines that have adjustable timing, certain conditions must be set before the timing can be checked or adjusted.
Answer B is wrong. The PCM never has full control of the timing.
Answer C is wrong.
Answer D is wrong.

Question #22
Answer A is wrong. Carbon buildup could cause high readings.
Answer B is wrong. Consistent low readings could be caused by a slipped timing chain.
Answer C is wrong. Low readings on adjacent cylinders usually indicates a blown head gasket.
Answer D is correct. A vacuum leak affects air-fuel mixture, not cylinder sealing.

Question #23
Answer A is wrong because B is also correct.
Answer B is wrong because A is also correct.
Answer C is correct. Either overheating or lack of lubricant will shorten turbo life.
Answer D is wrong.

Question #24
Answer A is wrong. Engine compression has no relation to turbocharger boost pressure.
Answer B is correct.
Answer C is wrong.
Answer D is wrong.

Question #25
Answer A is wrong. Pitting cannot be caused by the bearings.
Answer B is wrong. Excessive end play would damage only the bearings.
Answer C is correct.
Answer D is wrong. An overheated turbocharger would affect only the bearings.

Question #26
Answer A is wrong. A stethoscope cannot locate a plugged PCV valve.
Answer B is wrong. A vacuum gauge is not needed to diagnose a PCV system.
Answer C is wrong. Because Technician A and B are both wrong, C is wrong.
Answer D is correct.

Question #27
Answer A is wrong.
Answer B is wrong.
Answer C is correct. A visual inspection of all vacuum hose and electrical connections in addition to checking for related DTCs are essential diagnostic steps.
Answer D is wrong.

Question #28
Answer A is wrong. A bad ignition module would not affect only certain cylinders, unless it was a distributorless ignition system where the ignition module controls the coil of paired cylinders.
Answer B is correct. Multiport injected engines usually fire the injectors in groups (banks). In the example, cyl 1,3 and 5 are a bank. If one injector (#3) has less resistance, and the current is limited, the amperage flow will be greater for #3 leaving 1 and 5 with not enough to operate properly.
Answer C is wrong.
Answer D is wrong.

Question #29
Answer A is correct.
Answer B is wrong. You should check the EGR passages and control systems, as well as DTCs.
Answer C is wrong.
Answer D is wrong.

Question #30
Answer A is correct. A digital voltmeter (high impedance) can be used to monitor O_2 sensor voltage.
Answer B is wrong. An O_2 sensor cannot be checked with diode tester or an ohmmeter.
Answer C is wrong.
Answer D is wrong.

Question #31
Answer A is wrong. A proper cooling system test involves a pressure test.
Answer B is wrong. A proper cooling system test involves recovery tank inspection.
Answer C is wrong. A proper cooling system inspection involves a thermostat inspection.
Answer D is correct. The heater core is part of the heating, cooling, and ventilation (HVAC) system.

Question #32
Answer A is wrong. A cylinder misfire would cause high emissions.
Answer B is wrong. A lean condition would increase HC and O_2.
Answer C is wrong. A faulty fuel pressure regulator can cause an overrich condition, increasing CO and HC.
Answer D is correct. A power relay would not affect emissions.

Question #33
Answer A is correct.
Answer B is wrong. Nylon tubing cannot make sharp bends or turns.
Answer C is wrong.
Answer D is wrong.

Question #34
Answer A is wrong. High fuel pressure is not the least likely cause of poor fuel mileage.
Answer B is wrong. A disconnected regulator is not the least likely cause of poor fuel mileage.
Answer C is wrong. A partially plugged exhaust is not the least likely cause of poor fuel mileage.
Answer D is correct.

Question #35
Answer A is wrong. It is more common for the vacuum source to cause a problem than a blown fuse.
Answer B is correct.
Answer C is wrong. It is easier to check for a vacuum supply first.
Answer D is wrong. The exhaust system has nothing to do with this fault.

Question #36
Answer A is wrong.
Answer B is wrong.
Answer C is correct. The canister filter is a normal maintenance item. The filler cap must be checked for vacuum and pressure holds.
Answer D is wrong.

Question #37
Answer A is correct. Tight bushings can cause added drag on the starter.
Answer B is wrong. Worn brushes will not cause the starter to have low rpms.
Answer C is wrong.
Answer D is wrong.

Question #38
Answer A is wrong. Poor fuel mileage is a result of improper timing.
Answer B is wrong. If the belt has slipped enough, the engine won't start.
Answer C is correct. Vacuum reading would tend to be lower than normal.
Answer D is wrong. Improper timing can cause a lack of power.

Question #39
Answer A is wrong.
Answer B is wrong.
Answer C is correct. If the plunger is stuck open, it will be an equivalent of a vacuum leak, and will draw excessive crankcase vapors and possibly engine oil into the intake.
Answer D is wrong.

Question #40
Answer A is correct.
Answer B is wrong. The PCM might not be signaling the proper actuators.
Answer C is wrong. Lack of a vacuum signal could cause a failure.
Answer D is wrong. A solenoid failure could prevent the vapor from entering the engine intake.

Question #41
Answer A is correct. All resistance (carbon core) wires have a specification. Use an ohmmeter to check the wire, disconnected from the spark plug and distributor cap/coil.
Answer B is wrong. Spark plug wires always have some resistance.
Answer C is wrong.
Answer D is wrong.

Question #42
Answer A is wrong. Digital voltmeters can be used when working with PCMs.
Answer B is wrong. Most PCM inputs are low in voltage.
Answer C is correct. Most analog meters are of low impedance design, causing potential damage or incorrect reading in computer circuits.
Answer D is wrong. The oxygen sensor produces voltage from 500 millivolts to 1 volt.

Question #43
Answer A is wrong. An air manifold restriction will not cause a backfire on deceleration.
Answer B is correct.
Answer C is wrong. An exhaust manifold check valve will not cause a backfire on deceleration.
Answer D is wrong. An output pressure air pump will not cause a backfire on deceleration.

Question #44
Answer A is correct. A change in vacuum will either open or close the valve for tank vapor control to the canister.
Answer B is wrong. The valve does not have to be installed on the vehicle.
Answer C is wrong.
Answer D is wrong.

Question #45
Answer A is wrong. Components should not be replaced without using the flow chart.
Answer B is wrong. All steps in the flow chart must be followed.
Answer C is wrong.
Answer D is correct.

Question #46
Answer A is wrong because B is also correct.
Answer B is wrong because A is also correct.
Answer C is correct. Both are valid tests, eliminating either the secondary or primary ignition circuit as the source of the problem.
Answer D is wrong.

Question #47
Answer A is wrong because B is also correct.
Answer B is wrong because A is also correct.
Answer C is correct. The crank, cam, and distributor must be in proper relationship to a specific cylinder, usually number one, when reinstalling a distributor and setting timing.
Answer D is wrong.

Question #48
Answer A is wrong. Throttle bodies do not necessarily have to be removed to be cleaned.
Answer B is correct. These items need to be verified in addition to performing an idle relearn procedure and clearing any DTC as a result.
Answer C is wrong.
Answer D is wrong.

Question #49
Answer A is wrong. A rich fuel mixture would not cause knocking.
Answer B is wrong. A bad crank sensor would not cause knocking.
Answer C is wrong. High manifold vacuum would not cause knocking.
Answer D is correct. Restricted EGR would increase combustion temperatures, causing knock and NO_x.

Question #50
Answer A is wrong. It will save diagnostic time.
Answer B is wrong. It will show midproduction changes.
Answer C is correct.
Answer D is wrong. It will show year, make, and model information.

Question #51
Answer A is wrong. Constant high O_2 sensor voltage would not be the most likely cause.
Answer B is wrong. An EGR valve not seating would not be the most likely cause.
Answer C is wrong. A bad ignition module would not be the most likely cause.
Answer D is correct. This would allow vapors in the fuel canister to be purged at an idle, rather than at cruise.

Question #52
Answer A is wrong. Nylon hoses should always be inspected for kinks.
Answer B is wrong. Loose fittings should be inspected for fuel leakage.
Answer C is correct. Discoloration will not affect performance of the component.
Answer D is wrong. Nylon can be scratched easily and should be inspected.

Question #53
Answer A is wrong.
Answer B is wrong.
Answer C is correct. Any rpm drop not relative to the other cylinders indicates no power from that cylinder and can be caused by a fuel, ignition, or mechanical problem.
Answer D is wrong.

Question #54
Answer A is wrong. This would result in a no-start.
Answer B is wrong. Checking base timing at 2500 rpm would cause the timing to advance.
Answer C is wrong.
Answer D is correct.

Question #55
Answer A is wrong. Many maintenance-free batteries have neither filler nor translucent cases.
Answer B is correct. Some maintenance-free batteries have a translucent case that allows the electrolyte level to be seen; but this is only a general indication of the level, not an accurate check.
Answer C is wrong.
Answer D is wrong.

Question #56
Answer A is correct. System pinpoint testing is done *after* the complaint is verified.
Answer B is wrong. Always identify the conditions about the complaint.
Answer C is wrong. Road testing a vehicle can verify a complaint.
Answer D is wrong. The repair order contains information that needs to be reviewed.

Question #57
Answer A is wrong. Static electricity will harm a PCM, regardless of the battery cable connection.
Answer B is wrong. You should always ground yourself.
Answer C is wrong.
Answer D is correct.

Question #58
Answer A is wrong. The voltage drop on the positive cable should not exceed 0.5 volts.
Answer B is wrong. The voltage drop on the positive cable should not exceed 0.5 volts.
Answer C is wrong.
Answer D is correct.

Question #59
Answer A is wrong. Opens in the secondary can cause arc—through damaging windings.
Answer B is wrong. Uncorrected high voltage output requirements can overheat the coil.
Answer C is wrong. If a coil case is cracked, it is defective.
Answer D is correct.

Question #60
Answer A is wrong.
Answer B is wrong.
Answer C is correct. Either condition results in a rich mixture.
Answer D is wrong.

Question #61
Answer A is wrong. A restricted exhaust system will result in a continuous drop.
Answer B is wrong. A restricted exhaust system will result in a continuous drop.
Answer C is wrong. A restricted exhaust system will not cause fluctuations.
Answer D is correct.

Question #62
Answer A is correct. The fan will need to move more air when hotter.
Answer B is wrong. When the coil is cold, the orifice is closed.
Answer C is wrong.
Answer D is wrong.

Question #63
Answer A is wrong.
Answer B is wrong.
Answer C is correct. A fuel pressure test does several things. Vehicles with electric fuel pumps, the test confirms power, ground and pump integrity, in addition to verifying fuel pressure and volume. It is possible for a fuel injector to function correctly electrically, but not output the fuel properly.
Answer D is wrong.

Question #64
Answer A is correct. It is important to test the module for proper primary ignition control, especially with a no-start complaint. Some modules that fail do so under different conditions, for example, changes in temperature.
Answer B is wrong. An ignition module tester can only test the module's ability to switch on or off.
Answer C is wrong.
Answer D is wrong.

Question #65
Answer A is wrong.
Answer B is wrong.
Answer C is correct.
Answer D is wrong.

Question #66
Answer A is wrong. Distributor advance does not affect spark strength.
Answer B is wrong. Insulation does not affect spark strength.
Answer C is wrong. Low resistance will not cause weak spark.
Answer D is correct. Added resistance will reduce amount of voltage available to the plug.

Question #67
Answer A is wrong. The voltmeter should not read less than 9.6 volts.
Answer B is wrong. The voltmeter should not read less than 9.6 volts.
Answer C is correct.
Answer D is wrong. The voltmeter should not read less than 9.6 volts.

Question #68
Answer A is wrong.
Answer B is wrong.
Answer C is correct. Either condition will cause a rich mixture.
Answer D is wrong.

Question #69
Answer A is wrong. Using carburetor cleaner could be a fire hazard.
Answer B is wrong. Water cannot locate a vacuum leak.
Answer C is wrong. A stethoscope cannot pinpoint a leak.
Answer D is correct.

Question #70
Answer A is correct. A scope cannot be used for exhaust gas testing.
Answer B is wrong. A MAP sensor can be tested with an oscilloscope.
Answer C is wrong. A TPS sensor can be tested with an oscilloscope.
Answer D is wrong. A crank position sensor can be checked with an oscilloscope.

Question #71
Answer A is wrong. The fuel pump is not part of this test.
Answer B is wrong. Grounding the wire would result in a direct short.
Answer C is wrong. Grounding the wire would result in a direct short.
Answer D is correct. An approved spark tester will require the coil to put out approximately 25 kV, usually sufficient for all engines.

Question #72
Answer A is wrong. This checks battery voltage.
Answer B is wrong. This would check resistance.
Answer C is wrong.
Answer D is correct. To properly check for a draw, an ammeter needs to be used.

Question #73
Answer A is wrong.
Answer B is wrong.
Answer C is correct. A leak in the diaphragm will cause the fuel pump to not be able to draw the proper volume of fuel, or cause it to draw air. It must be able to build pressure, and maintain pressure.
Answer D is wrong.

Question #74
Answer A is wrong.
Answer B is wrong.
Answer C is correct. Any change in the delivery of the air-fuel charge will result in a performance or economy complaint.
Answer D is wrong.

Question #75
Answer A is wrong. If a circuit has excessive resistance, there would not be high current draw.
Answer B is wrong. A defective battery would not cause high current draw.
Answer C is correct.
Answer D is wrong. An ignition switch has no relationship to high current draw.

Question #76
Answer A is wrong. Both tools are not necessary.
Answer B is wrong. Accurate operation of the thermostat cannot be checked visually while installed.
Answer C is wrong.
Answer D is correct.

Question #77
Answer A is wrong. This is a normal condition.
Answer B is wrong. The O_2 sensor needs to be kept hot to function properly.
Answer C is wrong because both A and B are wrong.
Answer D is correct.

Question #78
Answer A is wrong. Low voltage will not harm the ignition switch or starter.
Answer B is wrong. Power to the ignition switch is fused.
Answer C is wrong. Because Technician A and B are both wrong, C is wrong.
Answer D is correct.

Question #79
Answer A is wrong. A specific gravity reading of 1.15 does not indicate a fully charged state.
Answer B is correct. If the battery gets too hot while recharging, it can be damaged. Refer to specifications for proper charge rate based on battery capacity, state of charge, temperature, and type of charger being used.
Answer C is wrong.
Answer D is wrong.

Question #80
Answer A is wrong.
Answer B is wrong.
Answer C is correct. This excites the field, causing the alternator to produce near full output. This is basically used as a "process of elimination" when diagnosing charging systems, as it separates the alternator and regulator circuits.
Answer D is wrong.

Question #81
Answer A is wrong. Battery voltage should not drop below 9.6 volts.
Answer B is wrong. Battery voltage should not drop below 9.6 volts.
Answer C is wrong.
Answer D is correct.

Question #82
Answer A is wrong.
Answer B is wrong.
Answer C is correct. Both A and B are correct. If there is a leak in an adjacent cylinder, look at what is "common" to both cylinders.
Answer D is wrong.

Question #83
Answer A is wrong.
Answer B is wrong.
Answer C is correct. Older vehicles that have manual idle mixture adjustments, CO is adjusted using a lean drop method, or, adjusting to a specific CO output level, using an exhaust gas analyzer. This applies to most vehicles prior to feedback control systems.
Answer D is wrong.

Question #84
Answer A is wrong.
Answer B is wrong.
Answer C is correct. Either condition increases fuel pressure, thus causing excess fuel consumption and higher than normal emissions.
Answer D is wrong.

Question #85
Answer A is wrong. The pickup coil should always be within specification.
Answer B is correct. An open pickup coil winding would show high or infinity on the ohmmeter.
Answer C is wrong. A high resistance reading would indicate an open circuit.
Answer D is wrong. Erratic readings while moving the pickup coil leads indicate a bad connection.

Question #86
Answer A is wrong. An EGR valve stuck closed can cause an engine to spark knock.
Answer B is wrong. Poor fuel quality can cause detonation.
Answer C is wrong. Spark plugs often cause detonation.
Answer D is correct.

Question #87
Answer A is correct. If the intake air is restricted, the volumetric efficiency is reduced.
Answer B is wrong. A stochiometric air-fuel mixture will not cause a lack of power.
Answer C is wrong.
Answer D is wrong.

Question #88
Answer A is wrong.
Answer B is wrong.
Answer C is correct. The gas cap is an integral part of the design of the fuel and emission control system, and calibrated for the application. Internal fuel tank pressures can be affected by the use of an improper cap, or a cap that is faulty.
Answer D is wrong.

Question #89
Answer A is wrong.
Answer B is wrong.
Answer C is correct. Voltage drop tests will confirm bad connections and corroded wires. These faults add additional "loads" to the circuit.
Answer D is wrong.

Question #90
Answer A is wrong. Not all fuel tank problems require replacement.
Answer B is correct.
Answer C is wrong.
Answer D is wrong.

Question #91
Answer A is wrong. The entry orifice does not control carbon monoxide and carbon dioxide.
Answer B is correct.
Answer C is wrong.
Answer D is wrong.

Question #92
Answer A is wrong.
Answer B is wrong.
Answer C is correct. Check for resistance problems using a voltage drop test. Corroded connections or damaged wires may not be revealed only by a visual inspection.
Answer D is wrong.

Question #93
Answer A is correct. Without the field wire, there will be no output.
Answer B is wrong.
Answer C is wrong. Other tests, such as power and ground voltage drop, battery, etc., need to be done before condemning the alternator.
Answer D is wrong.

Question #94
Answer A is wrong because B is also correct.
Answer B is wrong because A is also correct.
Answer C is correct. Either problem will result in the PCM not able to detect a change in engine load, or driver input.
Answer D is wrong.

Question #95
Answer A is correct.
Answer B is wrong. It is easier to listen to the injector at different speeds.
Answer C is wrong. Base timing has nothing to do with the fuel injectors.
Answer D is wrong. Checking the fuel pressure should be done after listening to the injectors.

Question #96
Answer A is wrong. The head gasket leak can be detected with a four-gas analyzer.
Answer B is correct. A graphing/DMM is used to monitor O_2 signals.
Answer C is wrong. The cylinder misfire can be detected.
Answer D is wrong. Inspection and maintenance program analysis is required.

Question #97
Answer A is wrong. Thirty seconds is too long. Depending on the manufacturer, fifteen seconds is usually the maximum.
Answer B is wrong. Never test a battery at the cold crank rating. The specification is ½ the CCA.
Answer C is wrong.
Answer D is correct.

Question #98
Answer A is correct.
Answer B is wrong. Connecting the gauge to a ported vacuum port will read vacuum off idle, not at idle.
Answer C is wrong.
Answer D is wrong.

Question #99
Answer A is wrong. The timing pointer and the timing indicator must be aligned.
Answer B is wrong. An indicator mark aids in the realignment of the distributor.
Answer C is correct. Follow procedures for the specific engine.
Answer D is wrong. When timing a distributor, point the rotor at the distributor cap terminal for the specified cylinder (usually #1).

Question #100
Answer A is wrong. A dirty oil filter has nothing to do with the turbocharger unit.
Answer B is correct.
Answer C is wrong. A dirty air filter will not result in oil consumption.
Answer D is wrong. A broken exhaust pipe will not cause oil consumption.

Question #101
Answer A is wrong. A shorted airflow sensor would not cause this.
Answer B is correct. The system may not recognize driver demand or wide open throttle command if the throttle position sensor is defective. It may or may not set a diagnostic trouble code (DTC).
Answer C is wrong.
Answer D is wrong.

Question #102
Answer A is wrong.
Answer B is wrong.
Answer C is correct. Any interruption of the primary input signals will not switch the module.
Answer D is wrong.

Question #103
Answer A is wrong. This detects pinging and is used to retard timing.
Answer B is wrong. Engine speed is used to calculate advance curve.
Answer C is correct. Power steering load input is used for idle compensation, not timing.
Answer D is wrong. Engine load is used to calculate timing advance.

Question #104
Answer A is wrong.
Answer B is wrong.
Answer C is correct. For example, the body controller may use data from the PCM to determine when to turn dashboard warning lights on, i.e., temperature. The transmission or climate control systems may also be tied to the powertrain control module.
Answer D is wrong.

Question #105
Answer A is wrong. Specific gravity should not vary by more than 0.050 between cells.
Answer B is wrong. Specific gravity should not vary by more than 0.050 between cells.
Answer C is correct.
Answer D is wrong. Specific gravity should not vary by more than 0.050 between cells.

Question #106
Answer A is wrong.
Answer B is wrong.
Answer C is correct. There are several methods used to retrieve trouble codes that are stored in memory, as well as dynamic self-test codes. The scan tools also may be able to perform more advanced tests or allow monitoring of data, an addition to code retrieval and definitions.
Answer D is wrong.

Question #107
Answer A is correct.
Answer B is wrong. There is no provision for vacuum actuation.
Answer C is wrong.
Answer D is wrong.

Question #108
Answer A is wrong, gas additives are not the best way to clean injectors.
Answer B is correct.
Answer C is wrong. Air pressure can damage the injector.
Answer D is wrong. Never use a brass wire brush.

Question #109
Answer A is wrong. A head gasket leak would allow air to escape to an adjacent cylinder or the radiator.
Answer B is wrong. No air should escape. It is a closed system.
Answer C is wrong.
Answer D is correct.

Question #110
Answer A is wrong. A MAP sensor is directly related to the air-fuel mixture.
Answer B is wrong. A defective MAP sensor can cause surging of the engine.
Answer C is wrong. If the MAP sensor is bad, it can cause a rich air-fuel mixture.
Answer D is correct.

Question #111
Answer A is wrong. Increase of ignition dwell helps to saturate the coil with low system voltage.
Answer B is wrong. The injectors will help slightly richen to increase idle speed.
Answer C is wrong. The idle speed will increase to help drive the alternator faster at idle speed.
Answer D is correct. The steering will not be affected.

Question #112
Answer A is wrong because B is also correct.
Answer B is wrong because A is also correct.
Answer C is correct. Both A and B are correct. The coolant temperature is used to determine when to change to closed loop during warmup, as richer mixtures are required.
Answer D is wrong.

Question #113
Answer A is correct. Use only a digital, high impedance voltmeter.
Answer B is wrong. A test light cannot test an O_2 sensor, nor should you try.
Answer C is wrong.
Answer D is wrong.

Question #114
Answer A is wrong. HC can be measured with a four-gas analyzer.
Answer B is wrong. CO can be measured with a four-gas analyzer.
Answer C is correct. NO_x is considered the "fifth" gas.
Answer D is wrong. Oxygen content can be measured with a four-gas analyzer.

Question #115
Answer A is wrong. Battery testers cannot be used to measure parasitic loads.
Answer B is correct. Connect the ammeter in series, so any current draw passes through the meter, to obtain a reading.
Answer C is wrong.
Answer D is wrong.

Question #116
Answer A is wrong because B is also correct.
Answer B is wrong because A is also correct.
Answer C is correct. The cold start injector is operated through a thermo-time circuit.
Answer D is wrong.

Question #117
Answer A is wrong. This is the proper way to adjust mechanical lifters.
Answer B is correct.
Answer C is wrong. The feeler gauge is the proper way to measure clearance.
Answer D is wrong. The piston should be at TDC so that both valves are not under load.

Question #118
Answer A is wrong. Leaking injectors would cause a rich O_2 sensor signal.
Answer B is correct. The air is not switching, causing O_2 sensor signal voltage to remain low. The system will over-richen to compensate, thus the high CO and lean exhaust DTC.
Answer C is wrong.
Answer D is wrong.

Question #119
Answer A is wrong. A 2.5 volt drop is excessive and the problem should be checked.
Answer B is correct. This test result is an accumulative total. There are several connections and components involved. Each will need to be tested independently to pinpoint the fault.
Answer C is wrong.
Answer D is wrong.

Question #120
Answer A is wrong. Is an OBD-II standard.
Answer B is wrong. Is an OBD-II standard.
Answer C is wrong. Is an OBD-II standard.
Answer D is correct. The amount is 1.5 times, not 4.

Question #121
Answer A is wrong because B is also correct.
Answer B is wrong because A is also correct.
Answer C is correct. A weak coil may provide enough spark to fire normally at an idle, but not enough to meet demands at cruise or when accelerating. If the problem existed long enough, there may be a DTC stored.
Answer D is wrong.

Question #122
Answer A is wrong. A bad PCM would affect both injectors.
Answer B is wrong. An open connection would result in the injector not working at all.
Answer C is correct.
Answer D is wrong. A low charging system voltage would affect both injectors.

Question #123
Answer A is wrong. The windings should always have resistance.
Answer B is correct.
Answer C is wrong. Faulty windings would not indicate the condition of the EGR valve.
Answer D is wrong. Faulty windings would not indicate the condition of the MAP sensor.

Question #124
Answer A is wrong. Service bulletins are useful for all technicians.
Answer B is correct.
Answer C is wrong. Service bulletins are for any information that will help the technician fix the vehicle in the shortest amount of time, and are not published for only recall campaigns.
Answer D is wrong. The bulletins are for all systems on the vehicle.

Question #125
Answer A is wrong.
Answer B is wrong.
Answer C is correct. Corrosion or added resistance can occur anywhere in the circuit. A thorough visual inspection and voltage drop tests will help locate problem areas.
Answer D is wrong.

Question #126
Answer A is wrong. Resistance is an approved method for checking a coolant sensor.
Answer B is wrong. Checking the return voltage is an approved method for checking a coolant sensor.
Answer C is wrong. The coolant temperature sensor can be placed in heated water while measuring the resistance.
Answer D is correct.

Question #127
Answer A is wrong. A visual inspection should always be performed first.
Answer B is correct.
Answer C is wrong.
Answer D is wrong.

Question #128
Answer A is correct.
Answer B is wrong. A rotor is located after a coil wire in the circuit.
Answer C is wrong. The primary ignition circuit checks as good.
Answer D is wrong. A diode problem would not cause this failure.

Question #129
Answer A is wrong. Cracked or disconnected hoses would result in vapor odor.
Answer B is correct. If the EVAP system purges vapors from the canister at idle, rough idle will result.
Answer C is wrong.
Answer D is wrong.

Question #130
Answer A is wrong.
Answer B is wrong.
Answer C is correct. Although O_2 sensors can be tested off the car, the scan tool is used to check O_2 operation by monitoring voltage and to check for related codes.
Answer D is wrong.

Question #131
Answer A is correct.
Answer B is wrong. Replacing the pulley does not require belt replacement unless the belt was damaged.
Answer C is wrong. Because Technician B is wrong, C is wrong.
Answer D is wrong. Because Technician A is correct, D is wrong.

Question #132
Answer A is wrong. The windings of the EVR should always have resistance whether the EVR is open or closed.
Answer B is wrong. Faulty EVR windings would not indicate the condition of the EGR.
Answer C is wrong. The windings should have some resistance, so the winding condition is relevant.
Answer D is correct.

Question #133
Answer A is wrong. Changing the cam sensor position will misrepresent cam position.
Answer B is correct. Some sensors allow for adjustment.
Answer C is wrong.
Answer D is wrong.

Question #134
Answer A is wrong. This is a lean condition.
Answer B is wrong. A DTC may be set.
Answer C is correct. This is not a rich condition.
Answer D is wrong. The O_2 sensor is operating properly.

Question #135
Answer A is wrong. A loose manifold would not cause the gauge to fluctuate.
Answer B is correct. Perform a cylindar leakage test to comfirm.
Answer C is wrong.
Answer D is wrong.

Question #136
Answer A is correct. This signal is for computed spark timing.
Answer B is wrong. The PCM uses this signal.
Answer C is wrong. The PCM uses this signal.
Answer D is wrong. The PCM uses this signal.

Question #137
Answer A is wrong.
Answer B is wrong.
Answer C is correct. As components wear, clearance is often increased beyond an allowable tolerance, thus creating a noise. It is important to determine the exact source of the noise before any disassembly, due to the noise "traveling," making it seem like it is coming from another area.
Answer D is wrong.

Question #138
Answer A is correct.
Answer B is wrong. There are no fuel pump diagnostic codes.
Answer C is wrong. It is easier to check the fuel pressure and volume first.
Answer D is wrong. Inspecting the fuel lines should only be done if they are suspect.

Question #139
Answer A is wrong.
Answer B is wrong.
Answer C is correct. For example, the body controller may use data from the PCM to determine when to turn dashboard warning lights on, i.e., temperature. The transmission or climate control systems may also be tied to the powertrain control module.
Answer D is wrong.

Question #140
Answer A is correct. The pulse width increases to richen the mixture.
Answer B is wrong. Short term fuel is subtracted or reduced as a result of a rich condition.
Answer C is wrong.
Answer D is wrong.

Question #141
Answer A is wrong because B is also correct.
Answer B is wrong because A is also correct.
Answer C is correct. Extending the IAC valve eliminates all incoming air, except through the throttle plates. If the throttle plates are adjusted, the angle has changed and the throttle position sensor needs to be adjusted.
Answer D is wrong.

Question #142
Answer A is wrong. B is correct
Answer B is wrong. A is correct
Answer C is correct. MAP sensors should hold vacuum. Digital MAP sensors produce a frequency.
Answer D is wrong.

Question #143
Answer A is wrong
Answer B is wrong
Answer C is wrong
Answer D is correct. The brake switch input is separate from the throttle position sensor input.

Glossary

Accelerator A control, usually foot operated, linked to the throttle valve of the carburetor.

Accelerator pump A pump in the carburetor, connected by linkage, to the accelerator pedal.

Accessory drive As in the belt driven accessories under the hood-fan, alternator, A/C, power steering, air injection pump.

Air fuel mixture The proportion of air and fuel supplied to the engine.

Analyzer Any device, such as an oscilloscope, having readout provisions used to troubleshoot a function or event as an aid in making proper repairs.

Automatic choke A system that positions the choke automatically.

Back pressure The excessive pressure buildup in an engine crankcase; The resistance of an exhaust system.

Battery A device used to store electrical energy in chemical form.

Battery cable Heavy wires connected to the battery for positive (hot) and negative (ground) leads.

Battery charger A device used to charge and recharge a battery.

Bearing A device having an inner and outer race with one or more rows of steel balls.

Catalytic converter An exhaust system component to reduce oxides of nitrogen (NO_x), hydrocarbon (HC), and carbon monoxide (CO).

Check valve A device that allows the flow of liquid or vapor in one direction and blocks it in the other direction.

Coil That part of the ignition system that provides high voltage for the spark plugs.

Cold cranking amperage The number of amperes that a fully charged battery will provide for 30 seconds without the terminal voltage dropping below 7.2 volts.

Combustion chamber The area above a piston, at top dead center, where combustion takes place.

Compression The process of squeezing a vapor into a smaller space.

Compression test Checking an engine for compression as a troubleshooting technique.

Computer A system capable of following instructions and to alter data in a desirable way in order to perform operations without human intervention.

Cooling system The radiator, hoses, heater core, and cooling jackets used to carry away engine heat and dissipate it in the surrounding air.

Cruise control A system of automatically maintaining preset vehicle speed over varying terrain.

Customer complaint The description of a problem provided by the customer, usually the driver of the vehicle.

Cylinder balance A dynamic test that shorts out the engine cylinders one at a time and compares the power loss in each to pinpoint weak cylinders.

Cylinder head That part that covers the cylinders and pistons.

Cylinder leakage test A test to determine how well a cylinder seals when the piston is at top dead center and the valves are closed.

Deck The flat mating surfaces of an engine block and head.

Dedicated ground There are many ground connections on an automobile, some are dedicated to a particular component or circuit.

Diaphragm A flexible rubber-like membrane.

Digital ohmmeter A device that sends a small amount of current into an isolated circuit and indicates the amount of resistance in a numerical readout.

Digital voltmeter A device that reads the difference in voltage pressure at two points of an electrical circuit in a numerical readout.

Driveability A term used for any problem or compliant the driver might encounter in the engine control system or transmission control system.

Driveability problem A problem or compliant encountered in the engine control system or transmission control system.

Drive belt The belt or belts used to drive the engine-mounted accessories off the crankshaft.

Dwell The degree of distributor shaft rotation while the points are closed.

EGR An abbreviation for exhaust gas recirculation valve. A valve that meters a small amount of exhaust gas into the intake manifold during light cruise conditions to lower combustion chamber temperatures and reduce the formation of nitrogen oxides.

EGR valve A valve that meters a small amount of exhaust gas into the intake manifold during light cruise conditions to lower combustion chamber temperatures and reduce the formation of nitrogen oxides (NO_x).

Electronic control A control device that is electrically or electronically actuated.

Electronics Pertaining to that branch of science dealing with the motion, emission, and behavior of currents of free electrons.

Emission A product, harmful or not, emitted to the atmosphere. Emissions are generally regarded as harmful.

Emission control Emission tests are tests done on the vehicle to determine how it compares to federal government standards for tail pipe emissions, crankcase emissions and evaporative emissions.

Emission control Components that are directly or indirectly responsible for reducing harmful emissions.

Emission test The use of calibrated equipment to determine the amount of emissions that are being released to the atmosphere.

Engine A prime mover. A device for converting chemical energy (fuel) to useable mechanical energy (motion).

Engine management system An electronic system that monitors, regulates, and adjusts, engine performance and conditions.

Engine manifold vacuum The vacuum signal taken directly off the intake manifold or below the throttle plate.

Engine oil A lubricant formulated and designated for use in an engine.

Exhaust The byproduct of combustion; the pipe from the muffler to the atmosphere.

Exhaust back pressure The pressure that develops in the exhaust system during normal operation. Two pounds is cause for concern.

Exhaust gas recirculation valve A valve that meters a small amount of exhaust gas into the intake manifold during light cruise conditions to lower combustion chamber temperatures and reduce the formation of nitrogen oxides.

Exhaust port An opening that allows the exhaust gases to escape.

F An abbreviation for Fahrenheit, a temperature measurement in the English scale.

Fail-safe A default mode designed into many operating systems that allows limited function of a system when a malfunction occurs. This is to protect the system or to allow the driver to move the vehicle to a safe area.

Fan blade A flat pitched part of a fan that moves air.

Fault code A numeric readout system used as an aid in troubleshooting procedures, information about functions or malfunctions of an electronic system.

Fuel contamination Any impurities in the fuel system.

Fuel injector Electrical or mechanical devices that meters fuel into the engine.

Fuel pressure The pressure of the fuel in a injected or non-injection fuel system.

Fuel pressure regulator A device that regulates fuel pressure. Fuel injected engines require pressure regulators because some fuel pumps develop over 100 psi (690 kPa).

Fuel pump An electrical or mechanical device that pumps fuel from the tank to the carburetor or injection system.

Fuel volatility A term used to determine how rapidly a fuel evaporates or burns.

Hand choke A choke that is cable controlled manually.

Head gasket A sealing material between the head and block.

Hot lines Special telephone or computer lines for information access for help with problem solving.

Idle speed The speed of an engine at idle with no load.

Ignition system The system that supplies the high voltage required to fire the spark plugs.

Ignition timing The interval, in crankshaft degrees of rotation, before top dead center that a spark plug fires.

Intake manifold That part of the engine that directs the air fuel mixture to the cylinders.

Intake port That part of a cylinder, having a valve, that allows the air fuel mixture to enter the combustion chamber.

Jumped A term often used in reference to timing chains, belts, or gears; meaning valve timing is off, or out of specifications. This happens when parts wear or loosen.

Jump start To aid starting by the use of an external power source, such as a battery or battery charger.

Keep alive memory A program in many computerized devices that retains fault code information and other information necessary for the operation of the system.

Key off battery drain A term used for parasitic drains. Electrical demands on the battery while the ignition key is in the off position.

Knock sensor A sensor that signals the engine control computer when detonation is detected retarding ignition timing.

LFT: Long-term fuel trim "Permanent" addition or subtraction of fuel assignment for fuel injected vehicles.

Lubrication The act of applying lubricant to fittings and other moving parts.

Magnaflux A dry, nondestructive magnetic test to check for cracks or flaws in iron or steel parts.

Main bearing The bearing that supports the crankshaft in the lower end of an engine.

Module A semi-conductor device designed to control various systems like, ignition, engine control, steering, suspension, brakes, transmissions, power windows, power seats, windshield wipers, brakes, traction control, and cruise control.

Oil A lubricant.

Oscilloscope An instrument that produces a visible image of one or more rapidly varying electrical quantities with respect to time and intensity.

Oxygen sensor A device located in the exhaust system close to the engine that reacts to the different amounts of oxygen present in the exhaust gases, and sends signals the engine control computer so it can maintain the proper air/fuel ratio.

PCV An abbreviation for positive crankcase ventilation. A metering device connecting the engine crankcase to engine vacuum, which allows burning of crankcase vapors to reduce harmful engine emissions.

PCV valve A metering device connecting the engine crankcase to engine vacuum, which allows burning of crankcase vapors to reduce harmful engine emissions.

Pings Unscheduled explosions in the combustion chamber. Also referred to as spark knock or detonation.

Positive crankcase ventilation valve A metering device connecting the engine crankcase to engine vacuum, which allows burning of crankcase vapors to reduce harmful engine emissions.

Power balance A dynamic test that shorts out one engine cylinder at a time and compares the power loss to pinpoint weak cylinders.

PSI (psi) Abbreviation for pounds per square inch. Used in the English system for air, vacuum, and fluid pressure measurements.

Rev limiter A device that limits the revolutions (turning speed) of a device or component.

Scanner A device, usually hand held, that accesses the electronic systems on vehicles to obtain fault codes and operating parameters. Can be used to simulate signals and to verify operation of systems on some vehicles.

Scan tool A tester used to recall trouble codes.

Seal A gasket-like material between two or more parts or a ring-like gasket around a shaft to prevent fluid or vapor leakage.

Secondary air insertion system Outside air that is pumped into the exhaust system and the catalytic converter to promote continued burning and chemical reactions that reduce harmful exhaust gas emissions.

Sensor An electrical sending unit device to monitor conditions for use in controlling systems by a computer.

Serpentine belt A wide flat belt with multiple grooves that winds through all of the engine accessory pulleys and drives them from the crankshaft.

Severe service Any vehicle service beyond average conditions, such as a taxi cab.

Servo A device that converts hydraulic pressure to mechanical movement.

SFT: Short-term fuel trim Corrects short term exhaust conditions and load.

Shroud A hood-like device used to direct air flow.

Speed limiter Usually a program in the engine control computer designed to limit the speed of the vehicle because of tire speed rating.

Supercharged A belt driven device that pumps air into the engine induction system at a pressure higher than atmospheric pressure.

Supercharger A belt driven device that pumps air into the engine induction system at a pressure much higher than atmospheric pressure.

Tail pipe The tube-like components that directs exhaust vapors from the outlet of the muffler to the atmosphere.

Technical information Information found in manufacturers manuals, bulletins, reports, text books, and other such sources.

Technical service bulletin Periodic information provided by the manufacturer relative to production changes and service tips.

Temperature The heat content of matter as measured on a thermometer.

Timing The spark delivery in relation to the piston position.

Trouble code Numbers generated by the diagnostic system that refer to certain troubleshooting procedures.

TSB An abbreviation for Technical Service Bulletins. TSBs are made available by product manufacturers to make known product improvements, problems found, fixes, revisions, and safety related concerns.

Turbocharger A device, driven by exhaust gases, that pumps air into the engine induction system at a pressure higher than atmospheric pressure.

Unmetered Matter, such as air, that is entering a controlled area without being measured by a management system.

Unmetered air Air that is entering a controlled area without being measured by the air management system.

Valve timing The opening and closing of valves in relation to crankshaft rotation.

V-belt A rubber-like V-shaped belt used to drive engine mounted accessories off the crankshaft pulley or an intermediate pulley.

Verify the repair To retest a system and/or test drive the vehicle after repairs are made.

Voltage A quantity of electrical force.

Voltage spikes Higher than normal voltage often caused by collapsing magnetic fields.

Wastegate A device on superchargers and turbochargers that limits the amount of pressure increase in the intake to safe design limits.

Water pump A device, usually engine driven, for circulating coolant in the cooling system.

Wrap A term used for the amount of contact a belt has on a pulley.

Notes